Fractured Empires

Christian Fourie

Fractured Empires

This is a work of fiction. Names, characters, places, and incidents are either the product of the author's imagination or are used fictitiously. Any resemblance to actual persons, living or dead, events, or locales is entirely coincidental. The geological and geopolitical contexts described are based on publicly available information. All operational details, characters, and incidents are entirely fictional.

ISBN: 978-99945-932-0-0
First published 2026
Self-published through Lulu Press

"Beware, the truth is often stranger than fiction."
Chrigi, Ranger of the San Clan

PROLOGUE
Windhoek, 2026

The Namibian sun was still low when the first message arrived.

Namibian President Nadi-Omunongo did not read it immediately. She stood at the tall window of her office, hands clasped behind her back, watching Windhoek wake up beneath the Khomas highlands. The city moved the way it always had, cautiously, patiently, unaware that somewhere far beyond the Atlantic horizon, a line had already been crossed.

She had been awake since four. Not from anxiety, but from an alertness that comes with thirty years of public life, the body learning, eventually, to treat sleep as optional. She had made tea. She had read the agricultural summary her aide had left on the corner of her desk, something deliberately small-stakes, cattle yields and water tables, a reminder that the country existed beyond the crises competing for her attention.

At seventy-three, she had learned that nations rarely fall when they expect to. They fracture quietly, under polite language and technical briefings, long before the shooting starts. She had watched it happen elsewhere. She had watched the people who thought they were immune discover that immunity was something you tested only once.

Behind her, the secure terminal chimed again.

She did not turn.

Namibia had just discovered it might be rich beyond imagining. Offshore fields in the Orange Basin, Venus, Mopane, billions of barrels sleeping beneath the Atlantic seabed. Enough oil to lift a desert nation into prosperity. Enough to turn it into a target. The geology reports had been on her desk for four months. She had read them so many times the margins were worn soft. She knew what they said. More importantly, she had spent four months thinking about what they meant, and the distance between those two things was where governments made their most expensive mistakes.

Total Energies was preparing its final investment decision. Galp and Shell were drilling day and night. For the first time, the Office of the President had pulled direct control over the sector. No diffusion. No intermediaries. No excuses, just Namibia first. Some of the cabinet had worried this signalled aggression. She had told them it signalled seriousness, and that the world had learned to confuse the two because seriousness from a small African nation was something it had not yet learned to process.

A female president. A predominantly female-led cabinet. A small country standing at the edge of a very old colonial game of stealing resources from the people who lived above them. She was not sentimental about this. Sentiment was for speeches. What she felt in the early morning, before the day organised itself into appointments and briefings, was something more like stubbornness, a determination not to let the same story run again with different names.

Nadi-Omunongo turned from the window as her aide entered.

"Madam President," he said carefully, "Elena Voss is ready. Caracas mission update."

Caracas.

Nine days earlier, American helicopters had dropped into the Venezuelan capital at two in the morning. Maduro had been dragged from Fort Tiuna and flown north. Washington called it stabilisation. Venezuela called it invasion. The rest of the world was still deciding what language to use, which meant the people who moved fastest would write the first draft.

And yesterday, Russia had answered with an Oreshnik hypersonic strike over western Ukraine.

Different continents. Different excuses and causes. Same war.

"Send her in," the President said.

She did not yet know that Namibia had already been drawn into the conflict, not by soldiers or sanctions, but by information. That was how it started now. Not with armies crossing borders, but with data moving through tunnels no one had officially authorised, connecting people who had never met, in cities that had no formal reason to be in contact.

By the time the world realised what Namibia had been given, it would already be too late to decide whether it should have been shared, buried, or destroyed.

The question would be who would survive its release.

Chapter 1: The Visitor from Caracas

Elena Voss arrived without an entourage.

No aides. No phone calls taken in the corridor. Just a thin folder under her arm and the faint smell of jet fuel clinging to her coat. She had come directly from the airport, stopping only long enough to change her shirt in a bathroom at Hosea Kutako. The choice was deliberate. She did not want to arrive looking composed. She wanted to arrive looking like someone who had not stopped moving, because that was the truth, and the President would know the difference.

President Nadi-Omunongo waited until the door closed.

"Elena," she said. "Please take a seat."

Voss did. Straight-backed. Alert. She had the look of someone who had already decided what mattered and discarded the rest. It was a quality the President had noticed the first time they met, three years earlier, when Voss had been a mid-ranking trade attaché with an uncommon habit of telling ministers things they did not want to hear without appearing to enjoy it. That quality had got her promoted twice and mistrusted by at least four ambassadors, which the President considered an excellent record.

"Tell me what Caracas looks like," the President said. "Not the headlines."

Voss paused before she answered. Not collecting her words, but checking them, the way you check a door handle before deciding whether to open it.

"Quiet," she said.

The word hung between them. "Not calm," she added. "Quiet like a city that's been warned."

The President nodded once. She had heard that silence before. In briefings. In border towns. In places that learned too late how history announced itself. The quiet that came not from peace but from the held breath before something happened. She had been in Harare in 2008. She had been in Tripoli in 2011, leaving two days before she would not have been able to leave at all. She knew exactly which kind of quiet Voss meant.

"Give me the details."

"The port at La Guaira is cratered. The airport runway won't take civilian traffic for at least six weeks. Fort Tiuna is rubble. Rodríguez is acting president, but authority has fragmented to the point where the title is mostly ceremonial. Militias answer to money, not ministries, and the money is moving in channels that have very little to do with the Venezuelan government."

"And the people?"

"Waiting," Voss said. "For proof that this was done for them, not to them. They're not in the streets yet. That's not resilience, that's calculation. They've seen what happens when you move too soon."

The President folded her hands. She let the silence run for a moment. There was a version of this conversation that moved directly to strategy. She had been in enough rooms to know that moving to strategy before you had fully understood the situation was how countries made decisions that looked sensible in the meeting and catastrophic in retrospect.

"And the oil?" she asked.

Voss didn't look at her notes. She had not opened the folder since sitting down. "It's already being discussed as if it no longer belongs to Venezuela. The language has shifted. It's not 'Venezuelan oil' in the conversations that matter. It's 'recoverable assets' and 'transitional energy infrastructure.' The vocabulary of acquisition dressed as the vocabulary of responsibility."

That earned a longer silence.

"By whom?" Nadi-Omunongo asked.

"Everyone who arrived before the dust settled," Voss said. "Chevron. Exxon. Advisers with no flags on their lapels and very well-tailored suits. They're not planning a recovery. They're planning continuity, which means the existing extraction infrastructure operates under new management with the same profit margins and a different flag on the letterhead."

"That was always the risk," the President said.

"No," Voss replied, carefully. "The risk is believing oil makes you sovereign."

She met the President's eyes.

"It makes you predictable."

The words landed. Not as accusation. As diagnosis. The President recognised the distinction because she had made it herself, in different rooms, about different countries, and she was acutely aware that her own country had just moved into the category Voss was describing.

"You were there," the President said. "When the helicopters came."

"Yes."

"What did you see?"

Voss was quiet. Not the careful quiet of someone choosing words, but the brief stillness of someone deciding how honest to be, and concluding that honesty was the only useful currency in this room.

"I saw how fast the world moves when it thinks it's running out of time," she said. "The helicopters were in Venezuelan airspace before most of the diplomatic community had been informed there was a situation. That's not an operation planned in weeks. That's an operation planned in years, and held in reserve until the moment the calculation changed. What frightened me wasn't the speed. It was the patience that preceded it."

"And who," the President asked quietly, "was it running from?"

Voss slid a single photograph across the desk.

The President did not reach for it immediately. She looked at it where it lay, practising the pause before picking up anything that someone had placed in front of her with the deliberateness of someone who had decided this object needed to be in her hands.

"What am I looking at?" she asked.

"Not what," Voss said. "When."

The President picked it up.

A timestamp. A data header. A routing path that shouldn't exist unless someone had wanted it to exist, and had wanted it to be findable, but not immediately. Not accidentally. Six minutes was what separated a coincidence from a confirmation.

Six minutes.

She set the photograph down and looked at it for another moment before looking up.

"And this involves us how?" she asked.

Voss exhaled, slow and controlled. It was the first time she had shown anything that might be called tension, and the President noted it precisely because it was so brief.

"Because someone in Caracas believed Namibia being non-aligned and neutral could make a difference," she said. "Not a military difference. Not a diplomatic one, in the formal sense. An informational difference. A country that might hold something without immediately selling it to the nearest interested party."

"That's an assumption," the President said.

"Yes," Voss agreed. "And assumptions are expensive and dangerous." She paused. "But they're also sometimes correct."

The room felt smaller now. Not literally, but the kind of smaller that came when the walls of a conversation had moved inward and the exit was no longer where you expected it.

"What do they want?" the President asked.

"To be believed," Voss said. "And protected long enough for the information to matter. Long enough to steer events away from a trajectory that ends badly for everyone, including the people who think they're controlling it."

The President rose and walked to the window. Windhoek spread below them, orderly, patient, unaware. She looked at it the way she often did when she needed to remember what she was actually responsible for. Not geopolitical calculations. Not leverage and timing. Forty-six cities and towns. A hundred schools opening this morning. A country going to work.

"And if they're wrong?" she asked.

Voss didn't answer immediately. "Then," she said, "we become a footnote to history." There was something in the way she said it that was neither dramatic nor resigned, simply factual. A footnote was a real possibility and should be acknowledged as one.

The President turned. "And if they're right?"

Voss met her gaze.

"Then we become something worse. A target."

Silence settled again, not awkward, not heavy. Decisive. The President had been in rooms where silence meant paralysis. This was not that. This was the silence of two people who had understood something at the same moment and were deciding what to do about it.

"Bring me everything," the President said at last. "But no intermediaries. No committees. Not yet."

Voss stood. "Yes, Madam President."

As she reached the door, Nadi-Omunongo spoke again.

"Elena."

Voss turned.

"If this is what you think it is," the President said, "history will pretend we had a choice."

Voss nodded once. "They always do."

The door closed.

The President remained at the window, holding the photograph between her fingers. Outside, the city continued as if untouched. The morning traffic had thickened along the Nelson Mandela road. Somewhere below she could hear the faint sound of a school bus. She stood there longer than she normally allowed herself, turning the photograph over once without looking at it.

Then she set it face-down on the desk and picked up her phone.

The first line had already been crossed. The question now was which lines remained, and who was watching to see if she walked toward them.

Chapter 2: The First Thread

Kyiv outskirts.

The basement smelled of damp concrete, burnt coffee, and overheated circuitry.

Three monitors lit the room. On the centre screen, the same footage looped again and again: a white streak tearing across the Ukrainian night sky, fast enough to erase sound itself. The Oreshnik strike had lasted seconds. The consequences were still unfolding. Andriy had watched the footage perhaps two hundred times in the past eighteen hours, not from morbidity but from the coder's instinct that repetition sometimes revealed what a single viewing concealed. He had not found anything new. What he had found was that the footage no longer frightened him, and that frightened him more.

Andriy Kovalenko, Spectro, to the few who still knew him that way, sat hunched forward, elbows on his knees, fingers moving between keyboards with the mechanical calm of exhaustion. Twenty-nine. Former cyber-defence, unit identifiers long since stripped. He hadn't slept since the strike. Neither had anyone else in the room, but exhaustion expressed itself differently in different people. In Andriy it had the quality of concentration, a narrowing of the world to the screen in front of him, every unnecessary thought stripped away by tiredness until only the work remained.

Behind him, Olena Marchenko stood with her arms folded, eyes fixed on the screen. She didn't pace. She didn't speak unless necessary. Stillness, she had learned, conserved energy, and energy was the one resource she could not acquire more of in the next twenty-four hours. She was thirty-four and had been doing versions of this work since she was twenty-seven, first officially, then in the unofficial way that people with skills too specific to leave behind found themselves continuing.

"Anything?" she asked.

"From Caracas, no," Andriy said. "From Windhoek, yes."

She turned.

"Elena Voss answered," he added. "No noise. Clean handshake. She knows what she's doing."

Olena nodded once. Something loosened slightly in her posture, not visible to anyone who didn't know her, but Andriy did. They had worked together for two years. He had learned her as he learned code, not by what was present but by what changed.

"Show her the spine."

Andriy opened a new window.

At first glance, the file looked thin. No logos. No dramatic headers. Just timestamps, routing paths, transaction fragments stripped of decoration. He had spent three days building it down to this: the minimum information required to make the connection undeniable without providing enough context to be immediately attributed.

It wasn't the volume that mattered.

It was alignment.

A financial transfer routed through three shells, each one incorporated in a different jurisdiction, none with any obvious connection to the others. A satellite access request logged outside normal scheduling windows, the kind of anomaly that disappeared into the noise of a system processing thousands of requests a day unless you knew what you were looking for. A secure-server handshake linking Caracas to Rostov-on-Don at a time when no officially sanctioned communication between the two cities should have been occurring.

Time-stamped: 01:47, January 3.

Six minutes before the first American helicopters crossed Venezuelan airspace.

Olena leaned closer. Andriy could feel her reading it, the quality of attention she brought to documents she was deciding whether to trust.

"Six minutes," she said.

"Enough," Andriy replied. "Too much."

"Coincidence?"

"There aren't coincidences at this altitude." He had said this before. He believed it the way he believed in mathematics, not as faith but as a description of how systems actually behaved when people thought no one was watching. Coincidence was what accidents looked like from a distance. From close enough, everything had a reason.

He trimmed the file further, removing everything that explained, leaving only what proved. Enough to be believed. Not enough to be traced back in a single move. He had learned this from watching other operations unravel, the ones that front-loaded their evidence and found that showing everything at once gave the other side too much to counter. You gave people what they needed to take the first step. The second step they took themselves.

He pushed it through the tunnel.

***

Windhoek

Elena Voss was awake before the phone vibrated. She didn't move immediately. She waited for the second signal that confirmed the channel was clean. This was habit, not paranoia, though the line between those two things had grown less distinct over the past two years. Then she picked it up.

Six minutes. The number settled in her chest like a weight, cold and precise. She had known it was coming since her conversation with the President that morning, but seeing it formatted as data, timestamped and routed and stripped of ambiguity, made it different. Knowledge was one thing. Evidence was another. Evidence required decisions.

Outside, Windhoek stirred. A call to prayer drifted faintly through the early air. Traffic thickened along the Nelson Mandela road. Somewhere nearby, a dog barked at nothing in particular. She had lived in this city for eight years and still experienced moments of genuine surprise at its ordinariness, the way it simply continued being a city regardless of what was happening inside the buildings her work led her into.

She typed carefully. *Received. Credible. I need three things: Chain-of-custody confirmation on the Caracas–Rostov link. Any identifiers tied to NordAzur, partial or masked. And your terms. What do you want from Namibia?*

She sent the message, then opened a blank document on her tablet and sat before writing anything.

The title came easily: Lessons from Caracas.

She hesitated, then added a single line beneath it, not a conclusion but a note to herself, the kind she would delete before anyone else read the document, marking where her thinking was at this particular moment: If the thread holds, the missile over Ukraine wasn't just another attack. It was a message sent in advance, to make sure the recipients were paying attention.

***

Caracas

The apartment was dark except for the glow of the laptop.

Daniela Ruiz sat cross-legged on the floor, back against the wall, listening to the building breathe around her. Generators hummed in the building across the street, cycling on and off with the regularity of a heartbeat. A siren wailed somewhere beyond the shuttered windows, then cut off abruptly, which in this city could mean the situation had been resolved or that the people with the siren had decided discretion was more useful than sound.

Twenty-one. Online, she was Luz. In the real world, she was careful, which was a different thing from being afraid, though both qualities had been sharpened by the same events. She had grown up in Caracas understanding that the distance between ordinary and dangerous was shorter than in other places. She had not found this tragic. She had found it clarifying.

The reply from Windhoek sat on the screen.

Received. Credible. She read it again. Someone had answered. Someone on the other side of the continent, in a country she had never visited, in a government she had no formal connection to, had read what they sent and decided it was worth responding to. She allowed herself a moment to feel what that meant before pushing it aside and typing.

Proof incoming.

NordAzur isn't a company. It's a corridor.

Our terms are simple: don't hand this back to the people who buried it.

A pause. Boots scraped on concrete somewhere below. Voices, male, at least two, too close and moving closer. Daniela closed the laptop halfway and waited, shoulders very still, breathing through her nose. The footsteps continued past. The voices faded.

She finished the message.

If this breaks the wrong way, no one involved survives long enough to explain.

She hit send and shut the screen.

In the darkness, she allowed herself a single small smile, not satisfaction, but the feeling of having done something that could not be undone. For most of her life she had been careful to avoid that feeling. In the past eighteen hours she had decided that being careful had its own costs, and that those costs were no longer ones she was willing to pay in silence.

Three continents were now connected by a thread no one had authorised. That was how the dangerous ones always began.

Chapter 3: Continuity
Washington, D.C.

The room was designed to be forgettable.

No flags. No portraits. No windows. Just a polished table, soundproof walls, and a clock that ran five minutes fast for reasons no one remembered but no one challenged. The furniture was the kind that suggested permanence without asserting it, functional and bland, entirely interchangeable with the furniture in thirty other rooms in thirty other buildings in this city where decisions that shaped other countries' futures were made in settings that looked like insurance offices.

Elliot Marsh arrived last. He took the empty seat without apology and set his phone face-down on the table. The screen went dark immediately. Whatever this meeting was, it would not leave a digital trail. He had attended meetings like this for twenty-two years. He had stopped finding them remarkable and started finding them informative in their ordinariness, in the way that people who believed they were invisible moved differently from people who knew they were being watched.

"Let's begin," said the woman at the head of the table. She had a name, but in rooms like this, names were rarely useful. "We're already late."

No one argued.

A map of the Caribbean lit the wall behind her. Venezuela pulsed faintly in red, the colour of something that required management rather than alarm. Marsh had seen this colour-coding before. Red meant attention. It did not mean emergency. Emergency was a different red, deeper, with a border.

"The transition is holding," she continued. "Ports are compromised but functional. Energy assets are secure. Markets are calm."

Marsh raised an eyebrow. "Markets don't stay calm," he said. "They wait. There's a difference, and the difference matters when you're deciding how much time you actually have."

A few smiles flickered around the table and died. He had not intended it as a joke.

"The Europeans?" someone asked.

"Concerned," the woman said. "Predictable. The Ukrainians have their own problem, which works in our favour in terms of where the international conversation is focused."

She gestured, and the map shifted. Western Ukraine glowed now, scarred by a thin white arc, clean and surgical and absolutely not random.

"The strike changed the temperature," Marsh said.

"It changed the narrative," another voice replied, from the far end of the table. "That matters more, at this stage."

Marsh leaned back. "Narratives fracture. Infrastructure doesn't. The strike was a demonstration. That means someone needed to demonstrate something, which means they felt they were losing the argument. People who are winning arguments don't fire hypersonic missiles at Ukraine."

Silence followed. The kind that meant someone had said the wrong true thing and the room was deciding whether to acknowledge it.

"Chevron and Exxon are in-country," the woman said at last. "Quietly. Advisory capacity only."

"Advisers don't move that fast," Marsh said. "Operators do. The distinction is legal, not logistical."

"They were prepared."

"Prepared for what, specifically?" he asked.

"For continuity."

Marsh smiled without humour. “Continuity is just collapse slowed down enough to feel responsible. I’m not objecting to the outcome. I’m asking about the timeline, because if the timeline was established before the operation was authorised, then we have a sequencing problem that will eventually become an accountability problem.”

No one responded. The woman tapped the table once, a gesture Marsh had come to understand as the equivalent of closing a door.

“Let’s stay focused. The objective is stability.”

“Stability for whom?” Marsh asked. He asked it genuinely. It was a question he had been asking in rooms like this for two decades and had never once received a direct answer to, which he found more informative than any direct answer would have been.

A beat.

“For the system,” she said.

There it was.

“And the leak?” Marsh continued. “The Caracas network. The data that moved before the helicopters did.”

“We’re monitoring.”

“Monitoring what, precisely?”

“Signals. Noise. Speculation.”

Marsh shook his head slowly, the way you shake your head when you are not surprised but are acknowledging that the conversation has arrived at a place you anticipated and are not pleased to find yourself standing in. “You don’t monitor leaks,” he said. “You cauterise them. Monitoring is what you do when you’re not sure the situation is serious. We passed that point six minutes before the helicopters crossed Venezuelan airspace.”

“We don’t have confirmation.”

"You don't need confirmation," he said. "You need timing. The confirmation will come from the people who already have it, and by the time it's confirmed through official channels, timing will no longer be available."

The woman studied him with the careful attention of someone deciding not whether he was right but whether his being right was a problem she needed to solve.

"You're suggesting action without attribution."

"I'm suggesting restraint is already a decision," Marsh replied. "Just not one we'll own, which is the part that concerns me."

Another map appeared. Southern Africa, the continent rendered in the clean flat colours of satellite cartography. Namibia sat in soft blue, the colour the system used for countries that had not yet been categorised, which was itself a categorisation.

"Why is this here?" someone asked.

"Because they haven't chosen yet," Marsh said. "And countries that haven't chosen yet are not neutral. They're available. That's a meaningfully different thing, and we should treat it as one."

"That makes them irrelevant."

"No," Marsh said. "It makes them dangerous. Irrelevant countries don't receive data transfers with six-minute timestamps. Irrelevant countries don't get chosen."

The woman frowned. "Namibia doesn't project power."

"They project legitimacy," Marsh replied. "That lasts longer. And in this particular situation, legitimacy is the resource that's most in demand and shortest in supply."

She considered this, filing it rather than dismissing it.

"They're not aligned."

"Exactly." He let the word sit. "Which means the first move they make defines them. Permanently."

Silence settled again, the kind where the room was doing calculation rather than avoiding it.

"Then we move carefully," she said finally. "No pressure. No visibility."

Marsh stood. He gathered nothing because he had brought nothing. "Pressure is already building," he said. "And visibility isn't optional anymore. Someone has already decided to listen, and listening is the beginning of a decision chain, not a pause in one." He buttoned his jacket. "The question isn't whether Namibia gets involved. It's whether we're in the room when they decide how."

He left without waiting for permission.

On the wall, the map of southern Africa remained illuminated long after the others dimmed. The soft blue of Namibia sat at the bottom of the continent like something that had not yet decided what colour it wanted to be.

***

Houston, Texas

The office on the thirty-second floor had a view of the ship channel that the senior vice president found useful on days like this, not for its beauty but for its utility. Tankers moved below in the slow, deliberate way of objects that could not be hurried. He found this reassuring. The physics of oil were the same as they had always been. It was only the politics that changed.

His secure phone rang. He answered on the second ring.

"We're in," said the voice. No preamble. No name.

"How fast?"

"Faster than we should be."

"And slower than someone else?"

"Yes."

"That's not acceptable."

"It never is."

He had been in this industry for thirty-one years. He had navigated three regime changes in two continents, one nationalisation that cost the company eight billion dollars, and two situations that had never been given names in any official document. He had learned that the worst outcomes came not from bad decisions but from good decisions made at the wrong speed, the moment when what you could do outran what you should do and nobody in the room had the standing to say so.

"What's the exposure if the timeline becomes public?" he asked.

The voice on the other end was quiet. "That depends on how much of it surfaces and in what order."

"Worst case?"

"Worst case," the voice said carefully, "is that the sequence tells a story we'd prefer people to reach on their own, over time, rather than read all at once."

He understood. He had spent thirty-one years becoming fluent in the language of managed disclosure.

"Manage the corridor," he said. "Whatever the corridor is. Narrow it."

The line went dead.

He stood with the phone still in his hand, looking at the tankers on the channel below. They were still moving at the same speed they always moved. The physics had not changed.

Risk, once shared, was no longer controllable. He had learned that the hard way, years earlier, in a country whose name he did not say aloud in offices anymore.

Chapter 4: Weight of Listening

Windhoek

The briefing room was smaller than most people expected. No raised dais. No flags behind the chairs. Just a circular table and walls lined with old photographs: independence rallies, drought years, ceasefire signatures faded by sun and time. The President had chosen this room deliberately, years ago, for exactly that reason. It was not a room where you felt powerful. It was a room where you felt responsible, and in her experience those two feelings led to different quality decisions.

President Nadi-Omunongo took her seat without ceremony.

Around her sat the core cabinet. Energy. Finance. Defence. Intelligence. Not advisers, decision-makers. She had insisted on this distinction from the first week of her government and had repeated it at sufficient intervals that it had settled into the culture of the room. These were not people who summarised other people's positions. These were people who held positions and were required to defend them.

Elena Voss stood, folder closed. The President had noticed that Voss had not opened it during the previous meeting either. This was not a performance of confidence. It was a statement about what kind of information she was dealing with, the kind that lived in your head because committing it to paper created vulnerabilities you could not control.

"Start with what we can't undo," the President said.

Voss nodded. "Someone has connected Caracas, Rostov-on-Don, and the Oreshnik strike," she said. "The connection isn't ideological. It's manufactured, built specifically to demonstrate that what looked like three separate events was one coordinated sequence, designed to escalate an already volatile situation toward confrontation at a scale none of the architects can fully control."

"Explain that," the Finance Minister said. Her name was Dr. Amina Iileka, and she had the precision of someone who had spent thirty years watching governments confuse correlation with causation in ways that cost their economies money.

"They moved information," Voss said. "Before anyone moved hardware. The financial transfers, the satellite access requests, the server handshakes, these existed before the helicopters flew. That means the operation was pre-authorised and pre-financed. It also means someone created a record, whether deliberately or carelessly. We're not sure which. The distinction matters."

A low murmur circled the table.

"And that implicates us how?" Defence asked. General Dlamini had the stillness of someone who had spent a career being the last person asked to solve problems, which meant he had learned to listen to the earlier parts of conversations very carefully.

"Because we answered," Voss said.

Silence followed. The kind that absorbed what had just been said and sat with it.

"Answered what?" Energy asked.

"A signal," Voss replied. "Not a request. A test. Whether Namibia was the kind of country that, when offered information this sensitive, would immediately sell it back to the nearest powerful buyer or sit with it long enough to understand what it meant."

The President folded her hands. "Go on."

"They believe Namibia will listen before it reacts. That we won't trade information for protection the way smaller countries are expected to. That our non-alignment is real rather than rhetorical." Voss paused.

"That belief is the basis on which they chose us. It's also the basis on which we'll be judged."

"And are they wrong?" Intelligence asked. Minister Kauru asked the question without inflection, not hostile, not supportive. Careful. He was the member of this cabinet most accustomed to the cost of being trusted with things that other people wanted to recover.

"No," the President said, before Voss could answer. "But belief creates obligation. That is the part of this we need to understand before we decide anything else."

Voss opened the folder for the first time.

Inside were not documents, but timelines. Transfers. Handshakes. Six-minute gaps that refused coincidence. She laid them on the table with the flat matter-of-factness of someone who had decided the information should speak for itself and did not need to be introduced.

"This," Voss said, "is what happens before wars become public."

Finance leaned back. Dr. Iileka looked at the timelines with the expression of someone doing arithmetic in her head that she did not like the result of. "If this is real, markets will move before governments do. They're probably already moving in ways that aren't visible yet. That means we have a window that's shorter than the political conversation assumes."

"They already are," Voss replied. "Just not visibly. The movements are designed to look routine."

Energy frowned. "And our oil?"

The room shifted. The question had been in the air since the meeting began.

"Our oil makes us legible," the President said quietly. "It tells the world we will matter. And it tells them when we will matter, and how much, and for how long. That is valuable information for people who are planning across a long horizon."

Heads turned.

"It also makes us a target," she continued, "not in the military sense, at least not yet. In the sense that countries with resources that other countries want are subject to a kind of attention that cannot be managed by simply being virtuous."

Defence exhaled slowly. "We don't have the capacity to absorb pressure from all sides simultaneously. We're a country of three million people with a defence budget calibrated for peacetime."

"No one does," the President said. "That's why they outsource it. They create the conditions under which a smaller country absorbs the pressure on their behalf, and they call it alignment."

"To whom?" Finance asked.

"To countries that haven't learned to say no yet," Intelligence replied. "Or countries that have said no and been made to understand the cost."

The President stood. She walked once around the end of the table, not pacing, thinking, using the movement the way she had always used it, to separate one thought from the next.

"We are not here to choose sides," she said. "We are here to choose sequence. What we do first determines what options remain for second. That is the only freedom we actually have in this situation, and I will not surrender it by moving before we understand what we're moving into."

She turned to Voss. "What happens if we ignore this?"

Voss didn't hesitate. "Someone else frames it. The information exists regardless of what we do. If we decline to be part of how it enters the world, someone else decides that, and we become characters in a story whose narrative we didn't influence."

"And if we expose it?"

"Someone accelerates. They move before the exposure can be used against them. We trigger the thing we were trying to prevent."

The President nodded. "Then we do neither." She looked at the table. "We listen. We verify. We decide what silence costs us, and we hold that number in mind before we decide what speech costs us."

Defence frowned. "Silence is already costing Ukraine. Has been for years."

"And noise destroyed Venezuela," the President replied. "I am not arguing that silence is safe. I am arguing that we need to know what we're choosing before we choose it." She met Dlamini's eyes. "Which is why we verify before we act."

No one argued.

"Compartmentalise," the President continued. "No committees. No written summaries that leave this floor. This conversation does not exist in any record until we decide it does."

"And understand this," she said. "The moment this leaves our hands in an uncontrolled way, we stop being observers. We stop having options. We become participants in someone else's timeline, and once you're in someone else's timeline, the exits are not where you expect them."

Energy spoke softly. "And if we're already past that point? If answering the signal already crossed that line?"

The President met her gaze. The question was honest and deserved an honest answer.

"Then history will pretend we volunteered," she said. "Which is why we are going to be very deliberate, from this moment forward, about what we actually choose."

***

Offshore Orange Basin, Namibia

The rig moved with the ocean, slow and patient.

Below the surface, two thousand metres down through water cold enough to kill and then into ancient stone, drills cut through geology that had not been disturbed since before human language existed. The pressure down there was a number that engineers converted into other numbers to make it manageable. The reality of it, the weight of the ocean above and the compressed heat below, was not a thing a human body could survive for the duration of a conscious thought.

In a control room lined with screens, an engineer frowned at a data spike that didn't belong. Not a large anomaly. The kind that appeared in systems this complex with sufficient regularity that most operators had learned to filter them. But this one had a pattern beneath the pattern, a regularity inside the irregularity that felt wrong in the way that professional instinct sometimes felt wrong before it could be articulated.

"Run that again," he said.

The numbers held.

Far away, in capital cities on three continents, decisions were being weighed. Governments were calculating. People who thought about oil in terms of leverage and timeline and strategic reserve were making assessments that would reshape the world above the ocean's surface.

Here, the earth simply responded to the drills the way it had always responded to disturbance, by revealing what it had been holding.

Chapter 5: Narrow Margins

Kyiv outskirts, January 12, 2026

The basement felt smaller. Not physically, but structurally, as if the air itself had thickened with the density of situations that had developed faster than anticipated.

Andriy stared at the new packet on his screen.

“It’s live,” he said.

Olena stepped closer. “Define.”

“They’re probing back. Not us directly, the corridor. But the corridor includes us, so the distinction is academic.”

She didn’t ask who. She already knew who. The question was what they knew and how much they had inferred from what they knew, which were two different problems requiring two different responses.

Traffic analysis bloomed across the monitor, not aggressive, not clumsy. Curious. Controlled. A sweep wide enough to look routine to anyone who wasn’t specifically watching for it, and narrow enough to be useful if it found what it was looking for. Andriy had seen this signature before. It was the kind of probe that state-level actors ran when they wanted information without wanting to be seen wanting information.

“They don’t know what they’re looking for yet,” Andriy said. “Just that something moved. Someone felt the displacement and is triangulating the source.”

“From us?”

“From somewhere we’re connected to. Which is the same problem.” He watched the traffic patterns shift slightly as the probe reached the edge of one of his mirrors and turned back. “They’re good,” he said, a professional observation. “They’re not going to stumble.”

“How long before they narrow it?”

"Hours. Maybe less if they throw more resources at it, which they will when this doesn't resolve quickly."

"And Windhoek?"

"Clean," he replied. "For now. I routed the handshake through four layers. It'll hold until it doesn't, which is how all of this holds."

Olena watched the scrolling lines for a moment longer. She was calculating something, he could tell by the way her stillness shifted into a slightly different quality of stillness, the kind that preceded a decision.

"Segment," she said.

Andriy split the network again, another layer, another mirror, another delay. Every move bought minutes. Minutes were survival. He had been in situations where survival was measured in hours and this had felt like luxury by comparison.

On the central screen, the Oreshnik footage still looped. He had stopped seeing it as footage. He saw it now as a timestamp, a before and after, the moment that had made this room and this work necessary in the way it was necessary.

"They used that strike as cover," Olena said quietly. Not for the first time. She had said it three times in eighteen hours, working something out.

"No," Andriy replied. "They used it as punctuation. Cover implies concealment. This message was meant to be seen. The strike said: the next phase has begun. Anyone who needed to understand that message understood it. Anyone who didn't was the audience, not the recipient."

Olena considered this. He could see her incorporating it, the slight adjustment of her posture that meant she had decided something.

He encrypted the next transfer manually. Slower. Harder to trace. The kind of caution that people who were certain they were safe did not bother with, which was precisely why it was worth bothering with.

"Send them the NordAzur trace," Olena said. "That narrows the circle. Gives Windhoek something to verify independently."

"It's already narrow." He hesitated only once before pressing send. Hesitation was not doubt. It was the last check before commitment, the moment when you asked whether you had thought of everything and answered honestly.

He pressed send.

***

Caracas

Daniela didn't go home, as a home had windows.

She had learned this the hard way, two weeks earlier, when the apartment she had been renting under a cousin's name had developed a problem with the window on the second floor, a problem that took the form of a person standing in the building across the street with equipment she recognised from a briefing she had attended when she was eighteen and briefly considering a career in a government that no longer existed in the form it had taken when she attended the briefing.

Instead, she moved between borrowed rooms, apartments owned by cousins of friends who asked no questions as long as she didn't stay long. She had become skilled at leaving no evidence of having been present. No dishes. No rearranged furniture. The kind of absence that looked like an absence rather than a departure.

The new message from Kyiv blinked onto her screen.

NordAzur corridor confirmed. Financial relay nodes: Marseille, Limassol, Dubai.

Her stomach tightened. She already knew Marseille. Had known it for seven months, since she had started pulling on the thread that had led to the thread that had led to this.

She had seen the invoices. Shell structures nested inside shell structures, each one incorporated in a different jurisdiction, the lawyers presumably congratulating themselves on the elegance of the construction while not asking too carefully about what was moving through it. Contracts written in language that never mentioned oil, only "continuity services", a phrase she had learned to read as a euphemism for the period between when a government stopped functioning and when a new arrangement started functioning, the hours and days when the actual decisions were made by people whose names did not appear in any official record.

She typed quickly. *Add this: shipment manifests tied to maritime insurance claims filed 48 hours before the strike.*

She hesitated.

The next sentence was the one she had been avoiding writing for six hours. She had drafted it four times and deleted it each time, not because it wasn't true but because writing it made it real in a different way than simply knowing it, the way that signing your name to something made it yours in a way that merely agreeing to it did not.

She deleted it again.

Then retyped it, because deleting it a fifth time would be a kind of cowardice she was no longer prepared for. It implicates continuity planners.

She hit send and closed the laptop.

A knock sounded down the corridor. Not her door, but close enough to make her very still for the three seconds it took to determine it was someone else's problem.

She exhaled and counted to ten. Then she gathered her things, systematically, the way she had learned to gather them, without urgency but without pausing, the practised efficiency of someone who had decided always to be ready to be somewhere else.

***

Washington

Elliot Marsh stood alone in his office, lights off, city glowing beyond the glass.

He had sent his assistant home two hours ago. He had read the same report three times, not because it changed but because he was looking for the thing in it he had not yet found, the element that would make the pattern coherent in a way that pointed to a specific next move rather than a general sense of deterioration.

His secure phone vibrated once.

"Say it," he answered.

"We're seeing movement," the voice said. "Fragmented, but intentional. Eastern Europe. Caribbean. Southern Africa."

Marsh exhaled slowly. "So someone chose Namibia." It was not a question. He had been half-expecting this since the meeting. Countries that hadn't chosen yet were, in his experience, the ones that other people chose for them, and the choosing was already underway.

"It appears so."

"And the leak? The Caracas material."

"Still unconfirmed in the official sense."

Marsh stared out at the Capitol dome, lit against the dark sky with the certainty of a building that had been there long enough to stop questioning whether it should be. "Leaks are always confirmed," he said. "Just not publicly, and not on timelines that are convenient for the people who'd prefer to manage the confirmation themselves."

"What do you recommend?"

He didn't answer immediately. He thought about Namibia. About what it meant that someone had chosen a country of three million people with a new offshore discovery and a president who, by all accounts, read everything herself and delegated nothing she didn't have to. That was a specific kind of choice. That was someone choosing the hardest path on purpose.

"Disrupt the corridor," he said at last. "Not the endpoints. Disrupting endpoints creates noise. You disrupt the middle, the relay points, and you create ambiguity about whether the information ever completed its journey."

"That risks exposure."

"Everything risks exposure," Marsh replied. "The question is what kind of exposure and on whose terms."

The line went dead.

He remained by the window. The city continued below with its usual indifference to the conversations happening in the lit offices above it. He had spent twenty-two years in those offices. He had spent most of that time believing that the conversations in them shaped the city below. He was less certain of that than he had been.

Continuity required friction. Friction generated heat. The question was whether the heat was controlled.

He was no longer certain it was.

***

Windhoek

Elena Voss watched the latest packet arrive.

Marseille. Limassol. Dubai. Insurance claims filed before the strike. Manifests tied to infrastructure that should not have been in motion until after the political situation was resolved.

Her jaw tightened. She forwarded nothing.

Instead, she printed the core pages, placed them in a plain envelope, and walked them personally to State House. No aide. No driver. She walked the four blocks in the morning air, the envelope under her arm, looking like a government employee on a routine errand, which was exactly what she was and exactly what she needed to appear to be.

Inside the President's office, she laid the envelope on the desk without opening it.

"They're accelerating," she said. "Not because we've done anything. Because of what we might do. The acceleration itself is information. It tells us we're holding something they'd prefer we didn't have."

The President didn't open the envelope yet. She looked at it the way she looked at things she was about to make decisions about, giving the object its due before letting its contents change the shape of the room.

"Because of us?" she asked.

"No," Voss replied. "Because of what we might do." She paused. "Which means they've already assessed us. They have an opinion about what Namibia does with things like this. The question is whether their opinion is correct."

The President considered that. The question was not rhetorical, and she did not treat it as one.

"And what might we do?" she asked quietly.

Voss met her gaze. She had been thinking about this answer since she walked the four blocks from the print room.

“Choose a side and stick to it,” she said. “Which is what they’re afraid of. Not because they fear Namibia. Because they fear the example.”

Chapter 6: Ripples across the Wire

Washington, D.C.

The Oval Office was a storm of activity, advisors milling, screens flickering with cable news feeds, the faint scent of fast-food wrappers lingering from a working lunch that had been served and mostly ignored.

President Jack Harvey sat behind the Resolute Desk, scrolling through his phone, thumbs hovering. He had been president for fourteen months. He had learned that the presidency was primarily a function of attention, whose you commanded and whose you lost, and that the mechanisms for managing attention had changed sufficiently in the past decade that the people who had briefed him on the job had not fully updated their models.

The Venezuela operation had been, by every metric that mattered to the people who measured such things, a success. Maduro in custody. Ports secured. Energy assets under reliable management. The headline cycle had run hard for seventy-two hours and was now beginning the natural process of domestication, the transformation of a geopolitical event into a political talking point, which was when it became really useful.

But the whispers were starting. Anonymous posts on X, leaks in European media about pre-raid chatter between Caracas and Moscow. Nothing solid, nothing that would survive a news cycle on its own merits, but enough to create the kind of friction that, left unaddressed, accumulated into narrative. His CIA Intel brief that morning had flagged it: Possible disinformation op. Russia pushing narratives to undermine the transition.

Harvey snorted. Time to set the record straight.

He tapped out the tweet:

“Venezuela is FREE at last! Maduro the Monster is in a New York jail where he belongs. Oil

flowing soon, CHEAPER GAS for Americans! No more Russian or Chinese puppets. The Fake News is spreading LIES about 'secret deals', ALL FAKE! We acted FAST and STRONG. America First! #MAGA"

He hit post. Within seconds, notifications exploded, the algorithm doing what algorithms did with content that combined declarative certainty with emotional charge.

But he wasn't done. He had learned that a single statement invited rebuttal. A second statement created a frame.

"To the losers in Moscow & Beijing: Hands OFF Venezuela! Your games are OVER. We know about your little 'missile gifts' to Maduro, ALL EXPOSED soon. Stay out or face CONSEQUENCES! Strong Borders, Strong America!"

He leaned back. Satisfied, almost. The CIA line in the brief lingered: Pre-raid chatter confirmed. Source credibility moderate. Moderate meant plausible. Let them deny it. The ripples would spread in ways that denial couldn't dam.

He did not think about the six-minute timestamp. It had not appeared in any brief he had read. He did not know it existed.

***

Moscow, Kremlin Press Room

The Russian Ministry of Defence briefing room was stark: fluorescent lights, a podium with the eagle emblem, reporters huddled in coats against the winter chill.

Major General Ivan Kohenkov stepped up, face impassive, uniform crisp. He had given versions of this briefing before, in this room and others, and had developed the quality of stillness that came from having said difficult things in public enough times that the difficulty had been processed in advance. He did not enjoy these briefings. He was good at them.

The Oreshnik strike was still fresh, footage replaying on loops worldwide, the hypersonic trail a white dagger over Lviv. The denials from Kyiv and Washington had been swift, performative, and entirely expected. Moscow stuck to the script because the script served its purpose regardless of who believed it. Belief was not the objective. Positioning was.

Harvey's tweet storm had complicated the morning. The reference to missile gifts was either a bluff based on fragmentary intelligence or confirmation that something had leaked. Kohenkov had received no briefing that told him which. He was delivering a statement prepared before he knew which, which was the only honest description of most public statements made in this building.

He cleared his throat.

"The Russian Federation categorically denies any involvement in alleged 'secret deals' with Venezuelan entities regarding military technology. Such claims are baseless provocations from the United States, aimed at justifying their illegal invasion and the blatant theft of sovereign resources."

A reporter from TASS raised a hand. "General, what about the Oreshnik strike? Was it connected to Venezuela?"

Kohenkov's eyes narrowed by a fraction. "The Oreshnik operation was a precise response to Ukrainian aggression against Russian sovereign territory, including the attempted strike on President Putin's residence on December 29, 2025. Any linkage to Venezuelan affairs is fabricated disinformation. Russia supports peaceful transitions but condemns US imperialism in Latin America."

The statement hit the wires immediately. Ripples turning to waves, each platform amplifying the signal according to its own particular logic, the sum of which was that the story was now bigger than any single actor's version of it.

In the car back to his office, Kohenkov sat with his phone in his hand and did not read the reactions. He looked out the window at Moscow's winter streets. The city looked the same as it had before the briefing. It always did. That was either reassuring or disturbing, depending on your relationship to continuity.

***

Windhoek, State House

Elena Voss watched the developments unfold on her phone in the antechamber outside the Security Committee room.

Harvey's tweets at 20:32 CAT. Russian denial minutes later. The speed of the sequence was the thing that caught her attention. Not the content, which was predictable, but the timing. Statements prepared in advance, released on a schedule. Which meant both sides had been waiting for a trigger.

The door opened. An aide nodded. "They're ready for you, Ambassador."

Inside, President Nadi-Omunongo sat at the head of the table, flanked by the ministers who needed to be in this room.

"Elena," the President said. "You've seen the latest?"

Voss nodded and projected her screen to the room's display. Harvey's tweets glowed large. Kohenkov's denial followed.

"The ripples have started," Voss said. "Either a partial leak or parallel intelligence reaching the same nodes. But the timing aligns with what our contacts provided. The Caracas-Rostov line is real. If we release the full package now, through the right channel, it could force a recalculation. Expose the backchannel structure. Make the escalation path too costly for any single actor to maintain."

Minister of Defence leaned forward. "And if it backfires? If the response isn't recalculation but escalation? We are not equipped to absorb retaliation from parties who feel we've damaged their interests."

"Nor are we equipped to remain silent indefinitely," countered the International Relations Minister. "Namibia's credibility at the AU and in the non-aligned movement rests on being a country that acts on principle. Silence, once you hold what we hold, is itself a position. It just isn't one we've chosen consciously."

Nadi-Omunongo studied the screen. She looked at both statements together, the American and the Russian, and read them not as political positions but as evidence of something she had been thinking about since her conversation with Voss that morning: the world was arranging itself around a confrontation that neither side had fully authorised and neither side could now fully control, and the people with the most accurate picture of how it had started were in this room.

"Conditions from the group in Caracas?" she asked.

Voss listed them. Exposure through neutral channels rather than attribution. Protection for the families involved. Some form of Namibian involvement in the process by which Venezuelan oil arrangements were renegotiated, to prevent the kind of extraction-by-other-means she had described to the President two days earlier.

"We can offer limited asylum," the President said. "Quietly, through SADC channels. That's a promise we can keep and one we can defend. Oil influence is harder: we propose an AU observer role in any transition framework. It's not power. But it's a seat, and seats matter more than they used to." She paused. "The disclosure itself routes anonymously to the UN. We are not the ones who release it. We are the ones who ensure that whoever receives it can verify it. That distinction matters for what comes after."

Elena hesitated for only a moment. The message, once sent, could not be recalled. She was authorising not just a disclosure but the beginning of a chain of consequences that she could model but not fully predict. Ten families. Ten lives whose futures changed the moment she pressed send.

She looked at the President. The President looked back without signalling direction, which was itself a signal, a reminder that this was Voss's decision to make and her responsibility to carry.

Green light. Leak via UN in 24 hours. Asylum: feasible for 10–15 families, coordinated through SADC. Oil: AU observer role to be proposed. Confirm?

***

Kyiv

Andriy saw the message ping. Olena read over his shoulder. She was silent for a moment.

"They're in," she said finally. Her voice was quiet in the way that things were quiet when they were real rather than when they were being performed. "Africa's committed."

Andriy stared at the screen. He had been working toward this for three weeks, through exhaustion and system failures and two separate moments when he had been certain the whole thing was going to collapse before it reached anyone who could use it. He did not feel relief exactly. He felt the flatness that came after the part you could control was over and the part you couldn't began.

"Are we sure about them?" he asked.

Olena looked at him. "We were sure about us once," she said. "That has to be worth something as a reference point."

***

Caracas

Daniela Ruiz, still in the dark apartment, felt her phone vibrate. She read the update.

A small spark of hope cutting through the fear was the closest description she had, though hope was not quite the right word. What she felt was more like the quality of forward motion after a long period of stillness, the sense that the thing she had been holding for months was finally going somewhere other than the place where she had found it.

Outside, the Hilux she had been monitoring for two days revved again in the distance. She registered it and did not act on it. There would be time for fear later. Right now there was only the work, and the work was moving.

Chapter 7: Pressure

Windhoek

The call came just after sunrise, when the light on the Khomas highlands was still the gold of early morning that Windhoek wore before the city fully woke.

"Madam President," the Finance Minister said, her voice measured but with a tightness underneath it that she was not quite managing to suppress. "We're seeing volatility."

"In what sector?" Nadi-Omunongo asked.

"All of them. Simultaneously, which is the part that concerns me. Simultaneous volatility across unrelated sectors doesn't happen naturally."

The President was at her desk. She had been reading the agricultural summary, her small-stakes morning anchor. She set it aside.

"Clarify."

"Sovereign spreads widened overnight, not dramatically, but enough to signal that someone has been talking to the right desks. Two scheduled energy infrastructure loans have been paused pending review, with no official explanation provided. Currency markets are jittery in a pattern that looks coordinated rather than reactive."

"Based on what trigger?"

"No official trigger. That's what I mean about coordination. This isn't a market reacting to news. This is a market that has received a signal through channels that don't leave paper trails."

She allowed herself a quiet exhale.

"What's the unofficial trigger?"

"A research note circulating in London. Anonymous. Three pages. Footnotes that look rigorous but reference things that can't easily be checked. It questions Namibia's institutional neutrality in emerging geopolitical conflicts. It doesn't accuse us of anything. It raises questions. Very responsible, very concerned questions."

There it was.

"Language?" she asked.

"Careful. Concerned. Responsible. The language of due diligence rather than attack. Which makes it harder to respond to, because responding to it validates the premise that there's something to be concerned about, and not responding allows the concern to develop in the absence of contrary information."

The President nodded once. This was a move she recognised, not because it had been used against her before but because she had watched it used against other governments and had spent twenty years thinking about how those governments had responded, which ones survived intact and which ones did not.

"Who amplified it?"

"Three major financial desks. Two think tanks. One Washington-based policy institute. The sequence suggests a cascade rather than independent judgment. They were waiting to amplify something. This was what they were given."

"Schedule the cabinet for this afternoon," she said. "And get me Voss. Not by phone. In person."

***

London

The report had no logo. It didn't need one.

Emerging Risk in Southern African Energy Governance.

Three pages. Footnotes. Plausible deniability baked into every sentence. The author, whoever they were, had the skill of the professional concern-raiser, a specific and valuable skill in the industry of managed perception: the ability to create anxiety without creating liability.

It never accused. It suggested. It asked whether small states, newly empowered by offshore discoveries, possessed the resilience to resist strategic manipulation from actors with longer timelines and greater resources. It cited Venezuela once, as a cautionary example. Ukraine twice, as a structural comparison. Namibia three times, always with the qualifier that no specific allegations were being made, that these were questions that prudent investors would naturally be asking.

The markets did the rest, as they always did.

***

Kyiv outskirts

"They've moved to narrative," Olena said.

Andriy scanned the financial ripple patterns on the monitor. He had not expected this to be the next move. He had expected something more direct.

"That's faster than I anticipated."

"They're not going to attack infrastructure first," she said. "That's visible. That's attributable. They attack confidence, because confidence doesn't leave forensic traces and it doesn't require authorisation at the level that kinetic action does. Someone in a London office can produce a document like this without it ever entering any official record."

On another screen, a wave of coordinated commentary surged across financial feeds. The language was consistent across sources that appeared to be independent. Governance maturity. Alignment risk. External influence.

"They're not naming us," Andriy said, reading through the commentary. "Not specifically. Not anywhere in the public record."

"They don't need to," Olena replied. "The financial community knows the geography. You say 'small state with offshore discovery, questions about neutrality' and the people who matter know exactly which small state you mean. The anonymity is procedural, not actual."

He looked at the Windhoek channel. The last message had been six hours ago. It was still clean.

"It's holding," he said.

"For now."

He thought about what it meant that they had moved this fast. It meant the release had been noticed before it was expected to be noticed. It meant someone with early visibility had moved immediately. It meant there was less time than they had calculated.

He typed a short message to Windhoek: *You're being tested publicly. Quietly. This is not spontaneous.*

***

Windhoek, Cabinet Room

The atmosphere had changed since the last meeting. Not panicked. But the quality of the alertness in the room had shifted from the alertness of people who were watching something develop to the alertness of people who were in something that was developing.

"They're manufacturing hesitation," Finance said. "Not collapse. The market moves are calibrated to create uncertainty without triggering the kind of alarm that would require official response. They want us to feel uncertain without being able to articulate what we're uncertain about."

"Why now?" Energy asked. "What changed in the past forty-eight hours that moved them to this?"

"Because we didn't react," Intelligence replied. "They ran the test of silence and we passed it, which told them the direct approach wasn't going to work. So they moved to indirect. They're trying to make the cost of not choosing a side feel more concrete than the cost of choosing one."

The President listened. She was watching her ministers think, which was one of the things she did in cabinet meetings and did not mention. She needed to know how they processed pressure, not just what conclusions they reached.

"And what do they want?" she asked.

"A statement," Defence said. "Alignment. Some kind of public signal that Namibia acknowledges the established framework. Not necessarily a choice of side, just a recognition that the framework exists and we operate within it."

"To whom?" the President asked.

No one answered immediately. The question was harder than it looked.

Voss entered, carrying a tablet with the latest monitoring. "They're escalating pressure through proxies," she said, taking her seat. "No direct contact from any official party. No warning through diplomatic channels. This is designed to be deniable from every direction, which means if we complain about it, we sound paranoid, and if we don't, it continues."

"Of course," the President said. Not with resignation. With recognition, the tone of someone encountering a problem they have studied before and find exactly as described.

"If we issue a statement, we validate the premise that there was something to issue a statement about. We confirm their framing. We give them the reaction they built the pressure to produce."

"And if we don't?" Finance asked.

"We absorb pressure. We demonstrate that our institutions function under conditions of financial uncertainty, which is itself information. It tells anyone watching that Namibia is not as fragile as the report suggested."

"How much can we absorb before the absorption itself becomes a story?" Finance asked. This was not defeatism. It was the question she was paid to ask.

The President looked at her. "Enough," she said. Then, more precisely: "More than they've calculated. That's the whole point of absorbing it."

A silence.

Voss stepped forward. "There's another development."

The room stilled in the way it stilled when the second piece of information arrived and everyone understood they were moving from one situation into a different one.

"NordAzur has gone dark."

"What does that mean in practical terms?" Energy asked.

"The corridor closed itself. The relay nodes went offline in sequence, starting with Dubai, then Limassol, then Marseille. In that order, which is the reverse of the money flow, which means someone is tidying from the outside in."

"Voluntarily?"

"No," Voss said. "Deliberately. There's a difference. Voluntary closure is disorganised. This has sequence and timing. Someone decided to close it and had the authority to close all three nodes within a six-hour window."

The implication settled over the table like temperature dropping.

"They're cleaning," Intelligence said.

"Yes."

"Can we prove anything before the cleaning is complete?" Defence asked.

"Fragments," Voss said. "We have what we already have. The window for acquiring more is closing."

The President rose slowly, the way she rose when she had made a decision and was about to announce it and was giving the room the courtesy of a moment to prepare.

"Then we don't chase the corridor," she said.

"Madam President?" Finance asked.

"We don't chase," she repeated. "Chasing confirms that we think the corridor is the evidence. It's not. The corridor is a symptom. We wait. If this was choreography, and the speed of the cleanup suggests it was, then someone just made an unscripted move. Choreography that gets disrupted reveals the choreographer." She looked directly at Voss. "We watch for who stumbles. Not for what they were running through."

***

Washington

Elliot Marsh reviewed the market response with the analyst who had been working the Namibia file.

"Effective," the analyst said.

"Measured," Marsh corrected. "Effective would have produced a visible reaction. Measured means they calibrated to produce pressure without one. The difference matters for what it tells us about Namibia."

"And Namibia?"

"Stable. For now. No public statement. No official response through any channel." The analyst consulted his screen. "No visible panic."

Marsh frowned. He had been in this business long enough to know that the absence of the expected reaction was data. Namibia's silence was not the silence of a country that had not received the message. It was the silence of a country that had received it and decided not to respond in the way the message anticipated.

"They should have reacted," the analyst said.

"Yes," Marsh replied softly. "They should have." He turned away from the screen. Something about this didn't fit the model, and things that didn't fit the model were the things that mattered. "Prepare contingency messaging."

"For what scenario?"

"For the one where restraint stops working. And for the one where it keeps working longer than anyone expected. Both scenarios require different messaging, and I'd rather have both ready than discover we've only prepared for the obvious one."

***

Caracas

Daniela watched the NordAzur nodes vanish from her interface.

One by one. Precisely. Like someone switching off lights in a building they were leaving, not in panic, in order. She had mapped this corridor over four months. She knew its shape. Watching it disappear with this kind of deliberate sequence felt like watching something she had built being demolished methodically by someone who understood its structure.

"They're burning it," she whispered.

Her encrypted channel blinked once. *Corridor closing. Prepare fall-back.*

"Fall-back to what?" she muttered. Fall-back assumed there was another position to fall back to. She stared at the screen for a moment, then began thinking through what she had before the corridor closed, what was preserved, what was already on its way somewhere that could not be recalled.

The Windhoek channel was still open. The timestamp was already in Namibia's hands. The insurance manifests, the relay nodes, the server handshake, all of it.

The corridor was closing. The information had already moved.

Outside, the city remained tense but ordinary. Inside her screen, a war of erasure was underway.

Pressure did not announce itself with sirens. It arrived as doubt, as hesitation, as polite questions about stability asked in rooms where the answer was already known.

And in Windhoek, a government that had chosen to listen now faced the cost of not speaking, while holding in its hands exactly the thing that would determine what that cost was.

Chapter 8: The Disclosure

Addis Ababa

The document did not arrive through official channels.

It was uploaded to a secure UN conflict-mediation archive by a user with credentials that technically existed but did not belong to anyone currently employed, the kind of account that a sufficiently motivated auditor would eventually find problematic, but that in the volume of traffic moving through that system on any given day would not be noticed until someone had a reason to look.

No announcement. No press release. No message sent to flag its presence.

Just a flagged entry in the system: *Supplemental Evidence, Venezuelan Transitional Risk Assessment.*

The duty officer in New York almost ignored it. The archive received dozens of supplemental filings every week, most of them genuine, some of them attempts to create a paper trail where none had previously existed, all of them theoretically important and practically buried under the weight of the institution's other concerns.

He read the first page.

Then the second.

Then he picked up his phone.

***

Windhoek

Elena Voss's phone vibrated once at 04:17.

Uploaded. No signature.

She did not reply. She lay still for a moment, phone face-down on the mattress, looking at the ceiling of the room she was sleeping in, which was not her apartment. She had not slept in her apartment in four days. The ceiling was unfamiliar and she was grateful for it, the same way she was grateful for anything that reminded her that the world she was operating in was not the normal world, and that treating it as the normal world was the kind of mistake you only made once.

She got up, dressed, and went outside.

The State House garden in the early hours had a stillness that was different from the stillness of a room. It was the stillness of things that existed before human activity and would continue after it. The jacarandas were motionless. The city behind the walls had not yet woken. The sky above Windhoek at 04:30 was a deep black with the first faint suggestion of differentiation at the eastern edge.

The President was already there.

This did not surprise Voss. She had begun to understand that Nadi-Omunongo kept the hours of someone who had learned that the early morning was the only part of the day that belonged entirely to her.

"It's done," she said.

The President looked at her for a moment before responding. "Did we leak it?"

"No."

"Did we stop it?"

"No."

The President allowed the faintest smile. It was the smile of someone who had prepared for a question and found the answer to be exactly what they had prepared for, which was not quite the same as relief but was adjacent to it.

"Good," she said.

They stood together for a moment in the garden. Neither of them spoke. There was something that needed to be said, eventually, about what came next, but both women understood that what came next could wait until the sun came up, and that standing here, in the last quiet before the day organised itself around consequences, was itself something worth not shortening.

***

New York

By the time Washington realised the file had been accessed six times in thirty minutes, it was too late to pretend it was accidental.

The duty officer had flagged it up. The flag had been received by someone who had forwarded it to someone who had escalated it to someone who had woken up the right people, and those people had spent forty minutes arguing about whether the file was real before acknowledging that whether it was real or not, the fact that six people had accessed it in thirty minutes meant it was already in the world.

The document did not accuse. It correlated. Time-stamped communications between Caracas and Rostov. Satellite vessel movements. Shell transfers through NordAzur's relay structure. Redacted appendices referencing third-party intelligence validation that the right reader would know how to interpret.

And then the annex. Twenty pages. Small.

Historic cross-references to 2014. Funding channels into what the document called Ukrainian civil resistance clusters. Sub-grants to security intermediaries whose names had long since disappeared from public records but would be very well known to anyone who had been in certain rooms during certain years.

The annex did not draw conclusions. It simply laid timelines side by side with the precision of someone who understood that the human mind drew its own conclusions from sequences and that those conclusions were more durable than any argument that could be made directly.

***

Langley

Elliot Marsh read it twice. Then once more, more slowly, the way you read something a third time when you are looking not for information but for the decision that generated it.

"Is it authentic?" he asked.

"Parts of it," the analyst replied.

"Which parts?"

"Enough. The core is clean. The NordAzur material is real, we recognise it. The timestamp data is real. The vessel movements are real. Some of the appendix is… interpretive. But interpretive in a way that's defensible."

Marsh closed the file. "They added Maidan."

"Yes."

"That wasn't in the original corridor. The Caracas material was about Venezuela. The Maidan annex is something else entirely."

He understood immediately, not all of it, but the shape of it. This wasn't a leak designed to wound Russia. It wasn't designed to expose the Venezuela operation. It was a leak designed to level the board. To make escalation symmetrically costly. To ensure that any country that decided to respond aggressively would be responding in a context where their own history was part of the same document.

"Who did this?" the analyst asked.

Marsh didn't answer, because the most accurate answer was both disturbing and clarifying at the same time: multiple actors, working from their own interests, had produced something that none of them could have produced alone. No fingerprints. That meant something more sophisticated than coordination. It meant restraint, the shared decision not to claim it.

"What do we do?" the analyst asked.

Marsh sat. He thought about Namibia. About a president who had chosen not to leak and not to stop it. He thought about what kind of country made that choice.

"We wait," he said. "We see who moves first. And we take notes, because what happens in the next seventy-two hours is going to tell us something about the world that we've been getting wrong for a while."

***

Moscow

Major General Volkov did not smile when he read it. He was not a man who smiled at information, finding the expression imprecise for the occasion.

He read it with the careful attention of someone identifying each element that required a response and each element that required nothing except patience.

"They included the annex," his aide said.

"Yes."

"Will we respond?"

"No."

"Not even denial?"

"Especially not denial." Volkov set the document down with the deliberate care of someone placing an object in a position they want to be able to find later. "Denial requires us to address the content. Addressing the content confirms its relevance. The annex has been in existence for twelve years. Let them explain 2014," he said quietly. "We have nothing to add."

"And Namibia?" the aide asked.

Volkov considered this. "A country that gives you what you need without asking for anything in return is either naïve or disciplined," he said. "Either way, it's valuable. We will watch."

***

London

The anonymous financial note was quietly amended within four hours of the document's first circulation. One paragraph removed without explanation. Two others softened to the point where the concern they had expressed now sounded more like general caution than specific warning.

The markets recalibrated. Namibian spreads stabilised over the course of a trading day.

No one admitted why. That was also, in its way, an admission.

***

Windhoek, Cabinet Room

"They've frozen," Intelligence reported. "Every major party. No escalation. No public rebuttal. No retaliation through any channel we're monitoring."

"In what sense frozen?" Finance asked.

"No movement of assets. No change in diplomatic posture. No narrative escalation on any front. The comment cycle has gone quiet in a way that doesn't happen naturally, which means someone has told someone else to stop talking."

"Which means?" Defence asked.

"They're calculating," the President answered. "Every major actor is now trying to determine whether acting makes their position better or worse, and discovering that in the current information environment, acting confirms whatever the last document suggested about them. So they're still."

She looked at Voss. "And NordAzur?"

"Dissolving. Corporate restructuring filings in three jurisdictions. Assets rerouted through new vehicles with no formal connection to the old structure. The people who ran it have moved on. It's being made to have never existed."

“So the corridor is gone,” Energy said.

“No,” Voss corrected. “It changed shape. The structure is gone. The relationships that ran through it still exist. The money that moved through it is already somewhere else. You can dissolve a company. You cannot dissolve the decisions that were made using it.”

Silence.

The President stood slowly. She had been thinking about what she wanted to say in this room since before sunrise, and had decided that the simpler the statement, the more accurately it would describe what had happened.

“We were supposed to pick a side,” she said. “We didn’t.” She looked around the table. “And now neither can they. Because picking a side requires a clean board, and the board is no longer clean. It has everyone’s fingerprints on it.”

She paused.

“We held. Now we watch what they do with the fact that we held.”

***

Caracas

Daniela watched the global feeds on her laptop screen, sitting in the third borrowed room in as many days.

No hysteria. No missile launches. No retaliatory sanctions. Just analysis panels struggling to connect dots that were already connected, intelligence reporters hedging their language in the way reporters hedged when they knew more than they were allowed to say, and a peculiar quiet spreading across the channels that usually carried noise.

Her secure channel blinked. *Exposure achieved. Casualties: none. Yet.*

“They didn’t burn,” she whispered.

"No," came the reply from Kyiv. *"They blinked."*

She sat with that for a moment. She had been prepared for a longer fight. She had not been prepared for this, for the sound of very large things becoming very still.

She was not naïve enough to think it was over. But she had learned, at twenty-one, that the spaces between the fire were where you made yourself ready for the next thing.

***

Kyiv outskirts

Andriy removed his glasses and rubbed his eyes.

He had been awake for thirty-one hours. The coffee had stopped working somewhere around the twentieth, and he had continued on the fuel of situations that refused to resolve, the engine of pure forward motion that kicked in when sleep stopped being an option.

"Do you realise what just happened?" he asked.

Olena nodded. She was sitting on the edge of the table, which she only did when she was very tired, using the furniture the way people used furniture when they no longer cared about the impression they made.

"We showed everyone's fingerprints," she said. "And no one can escalate without confirming them. The moment anyone acts aggressively, they're acting in the context of a public record that connects their actions to a much longer history. The cost of escalation just went up by everything that's now in that document."

"Which means," she said, "for now, deterrence."

He leaned back in his chair. "Namibia held."

"Yes."

"And the Americans?"

"Embarrassed. Not damaged. There's a difference, and they'll be calculating the difference for a while."

"And the British?"

"Quiet."

"And the Israelis?" he asked.

Olena considered the screen. "Watching. They always watch first. It tells them more."

He put his glasses back on.

The world did not explode. It paused. Which, in geopolitics, was much rarer and considerably harder to engineer. Missiles were still armed. Markets still moved. Intelligence services still ran operations whose existence would be denied in any public forum.

But for forty-eight hours, the temperature dropped, because exposure without accusation was a weapon that no one had a clean counter to. It did not require the accuser to be clean. It required only that the record exist and that the record be verifiable.

And somewhere between Windhoek, Kyiv, and Caracas, a coalition of actors with no formal relationship and no institutional backing had forced the largest powers in the world to reconsider their next move.

Not by strength.

By timing. And by the courage of people who decided that the truth, released carefully and without ownership, was still the most disruptive thing you could put into a world that had arranged itself around managing it.

Chapter 9: Blowback

Washington

The first mistake was calling it a "foreign disinformation packet." The second was believing that would be enough.

By mid-morning, three Senate offices had requested classified briefings.

Two former CIA directors had appeared on television, carefully neutral, using words like *context*, *historic complexity*, and *misinterpretation of archived operational frameworks*.

Which meant: yes, the annex was real, but not all of it, just enough.

Elliot Marsh stood in a secure conference room in Langley, staring at a projection of the UN upload. "Who authorized cross-theatre exposure?" he demanded.

No one answered. Because no one had.

The Maidan material had not been part of any current active file. It had been sealed under compartmentalized historic operations. It should not have been accessible to Venezuelan or African mediation channels. And yet it was.

"Internal breach?" someone offered. Marsh shook his head slowly. "No."

"Then what?"

"Someone used our own validation trails against us."

A younger analyst swallowed. "You think this was Russia or China?"

"No."

"Ukraine?"

"No."

"Then who?"

Marsh stared at the screen again. "That's the problem, we do not know yet."

London, Vauxhall Cross

MI6 did not panic, it restructured with three desks quietly reassigned. Two liaison cables were recalled.

An internal memo circulated: *Minimize commentary. Avoid historical narrative engagement. Focus on present-day stabilization frameworks.*

Translated: Do not discuss 2014. Do not validate the annex. Do not attack it either.

Markets had stabilized. Namibia had not been sanctioned.

There was no clean way to retaliate without drawing more attention, and the British were very good at silence when silence protect leverage.

***

Tel Aviv

Mossad's analysis division was less restrained.

"They've neutralized alignment pressure," one officer said.

"How?" The Director ask.

"By distributing culpability."

Screens displayed timelines: Caracas. Rostov. Maidan funding channels. Shell entities.

Everything placed side by side without editorial comment.

"It's elegant," another admitted reluctantly.

"And dangerous."

"Why dangerous?"

"Because it erodes moral positioning."

Silence.

The director spoke quietly. "Track Namibia."

"Hostile?"

"Yes."

"Then why?"

"Because they refused to behave predictably."

***

Moscow

The Kremlin did not celebrate. It recalibrated.

Volkov briefed behind closed doors. “The annex shifts perception of involvement,” he said.

“Does it absolve us?” a senior official asked.

“Not totally.”

“Then what does it do?”

“It complicates escalation.”

The room absorbed that. “If NATO moves aggressively now,” Volkov continued, “they validate the narrative of reactive containment.”

“And if they do nothing?”

“They appear constrained.”

“So we gain?”

“For now.”

“And Ukraine?”

Volkov paused. “Ukraine fractures internally if the neo-nationalists attempt to weaponize this.” He did not mention Pavel, as he did not need to.

***

Kyiv

The far-right forums ignited first. “Globalist betrayal.” “Foreign interference.” “Proof of Russian manipulation.”

The Banderite splinter cells attempted to reframe the leak as evidence of Russian treachery rather than CIA overreach.

But the annex did not support them cleanly. It showed funding. Not endorsement, but manipulation. No clear link. The nuance undercut extremism, which made them angrier.

Andriy watched the chatter scroll. “They’re losing control of the narrative,” he said.

Olena nodded. “Which means they’ll try something kinetic.”

“Against us?”

“Against someone.”

***

Caracas

Rodríguez held a press conference.

“The Bolivarian Republic categorically denies all foreign conspiracy theories regarding oil exchange for military hardware,” she declared.

But she did not deny NordAzur by name, or mention Rostov.

She did not accuse Namibia, because accusing Namibia would legitimize it. And legitimizing Namibia meant elevating a small African state into global arbitration. Which was unacceptable.

Daniela watched the feed. “They’re scared,” she whispered.

“Of what?” the Kyiv channel asked.

“Of losing the narrative.”

***

Windhoek, Cabinet Security Committee

The mood was sober and reserved.

Intelligence set her folder down without opening it. “They are probing.”

“On which fronts?” the President asked.

“All of them, simultaneously. Cyber, nothing destructive yet, more like someone testing which doors are locked. Financial, through third-party channels, trying to map our exposure. And diplomatic, through back channels we weren’t supposed to notice.”

Finance leaned forward. “The IMF meeting in Nairobi last week.”

“Among others.”

"I thought that felt off," Finance said. "The questions weren't about us. They were about who else we talk to."

"Exactly." Intelligence paused. "They're not after information. They're after a picture of our relationships."

The President listened without expression. "Are we targeted directly?"

"Not yet. What we're seeing is more indirect. Reputation pressure, with quiet suggestions in certain circles that our governance framework is untested at this level of resource wealth. Soft power isolation, nothing overt, just the right people not returning calls. And investor inquiries about political stability that are a little too coordinated to be coincidental."

Energy cleared her throat. "The oil majors?"

"Watching. Total had a call with our liaison last Thursday that was, by all accounts, deliberately inconclusive."

"Meaning they're waiting to see which way the wind blows before they commit."

"Meaning they've been asked to wait," Intelligence said. "There's a difference."

A brief silence settled over the table.

The President leaned back. "Then we have leverage. If they needed us marginalised before the investment decisions were made, they would have moved faster."

Voss spoke quietly from the end of the table. "They don't want to escalate publicly. A public confrontation gives us a stage, and they know we'd use it. So they're working the edges, testing where we flinch, looking for the threshold below which we'll absorb pressure without responding."

"And our response?" the President asked.

"Hold. Stay boring." Voss turned her pen over once. "Stay procedural. No speeches, no press statements that sound defensive, no emergency meetings that leak. The moment we look rattled, the investor hesitation becomes self-fulfilling. Right now our biggest asset is that we haven't blinked."

Defence had been quiet. "And if the external pressure doesn't find purchase and they try to fracture us internally instead? Someone in this room, someone in the civil service, someone who needs money or needs a favour?"

The question sat there.

The President's voice, when she answered, had cooled by several degrees. "Then they misunderstand Namibia. And we will take some time to explain it to them."

***

Langley

Marsh finally received what he'd been waiting for, a forensic breakdown of the upload. Source routing and encryption layering. Dead drops.

He studied it for a long time, then slowly he exhaled.

"This wasn't a hostile state."

"What do you mean?" the analyst asked.

"No fingerprints of Russian doctrine. No Chinese packet signatures. No Israeli digital watermarking. And not our own style."

"Then who?"

He tapped the screen where three access paths intersected. "Three mid-tier nodes."

"Which are?"

"Kyiv civilian cyber cell and Caracas opposition cluster."

"And…?"

"Windhoek."

The analyst blinked in disbelieve. "Namibia?"

"Yes."

"That's impossible."

Marsh shook his head slowly. "No, it's new."

Silence filled the room.

"So what do we do?" the analyst asked.

Marsh looked tired for the first time. "We don't overreact."

"That's it?"

"If we punish Namibia, we validate them."

"And if we ignore it?"

"We lose narrative control."

"Then what?"

He closed the file. "We wait for them to make a mistake."

***

Odessa

Pavel Sokolov's body was found near the docks, shot twice, with no insignia or traces left behind.

The neo-nationalist channels blamed Russian infiltration and Russian channels blamed Ukrainian radicals. No one mentioned the leak, as the fractures deepened quietly.

***

Windhoek

The President look at the data. "No missiles, no sanctions and no condemnations." she said softly.

Voss stood beside her. "They're recalculating."

"And the agencies?"

"Fracturing."

"And the war?"

"Paused."

The President nodded once. "We bought time."

"For what?" Voss asked.

The President did not answer immediately. "For something better than just mere survival." She turned away from the glass. "Or something much worse."

Outside, Windhoek slept peacefully.

In Washington, Moscow, London, and Tel Aviv, lights stayed on. Not because of bombs, but because of doubt, and doubt, in the architecture of empires, creates fractures.

Chapter 10: The Manufactured Spark

Washington

They did not call it retaliation. They called it stabilization.

A small inter-agency task cell was formed under a neutral title: *Strategic Exposure Containment Group*

Elliot Marsh did not lead it. He observed it. Which meant it was political.

"We isolate Namibia," one State Department official suggested.

"Too obvious," someone replied.

"Economic leverage?"

"Markets stabilized. Hard to justify."

"Cyber pressure?"

"Risky. If traced, it legitimizes their role."

Silence. Then a quieter voice: "Destabilize the coalition."

"How?"

"Make the leak look manipulated."

Marsh leaned forward slightly. "You're suggesting discrediting the source?"

"Yes."

"And how would you do that?"

"Insert conflicting data into the ecosystem. Seed contradictory forensic trails. Suggest Russian fabrication of Maidan annex."

"Without evidence?" Marsh asked.

"With ambiguity and gaslighting the facts."

He didn't respond, but he understood the direction. If you cannot deny the document, corrupt and gaslight the conversation around it.

***

London

MI6 ran parallel an analysis, as they did not coordinate openly with Langley, but the logic was similar.

"Namibia must not become a precedent," a senior analyst said. "If small states begin arbitrating intelligence disputes, we lose centrality."

A proposal circulated: Activate dormant media relationships. Question the technical authenticity of the annex. Emphasize Russian incentive to distort Maidan history. No overt attack, just create doubt.

***

Tel Aviv

Mossad's approach was different. No media campaigns. No narrative shaping. They probed digitally.

Low-level penetration attempts against Windhoek's secure communications. Not to breach, just to map. The message was subtle: We see you.

***

Windhoek, Parliament

The internal pressure began from an unexpected place.

Opposition MP Johannes Kautara stood and asked the question live on record: "Has the Namibian government involved itself in foreign intelligence dissemination without parliamentary oversight?"

The chamber shifted. The President did not flinch. "We have not engaged in unauthorized foreign intelligence operations," she replied calmly.

"Have we transmitted sensitive material to international bodies?" Kautara ask.

"We have supported peace mechanisms within established diplomatic frameworks." The President answered,

It was precise, true, and incomplete.

Outside Parliament, investors watched. Not panicked, but curious, as curiosity is the seed of volatility.

***

Kyiv

The neo-nationalist channels did not let the annex die. They reframed it.

"If the CIA funded us, then we were justified. If the West manipulated Maidan, then sovereignty demands cleansing."

A rally was announced. Not large but loud.

Olena watched the chatter carefully. "They're trying to force a domestic rupture."

Andriy nodded. "If NATO looks unstable, Russia gains, and if Russia overreacts, NATO regains moral ground. Which means someone will try to trigger something undeniable."

***

Caracas

The explosion was not large, as it did not destroy infrastructure. It destroyed timing.

A small blast at a Venezuelan oil transit facility near Puerto La Cruz, with minimal casualties, but high visibility.

Within minutes, anonymous channels pushed the claim:

"Retaliation for oil betrayal."

Russian accounts blamed CIA sabotage. Western accounts blamed pro-Russian militias. Caracas blamed "foreign destabilization actors."

The truth was more complicated.

Because the explosive signature did not match Russian doctrine, nor CIA procurement lines.

It was hybrid. Deliberately ambiguous.

***

Langley

Marsh stared at the forensic snapshot for a long time before he spoke.

"Did we do this?"

The analyst chose his words carefully. "Not through any channel I have visibility on."

"That's not what I asked."

"No, sir. We didn't do this."

Marsh looked up. "Moscow?"

"Unlikely. The signature doesn't fit their current operational tempo, and frankly it's too clumsy for what they've been running lately. They've been disciplined."

"Then who?" Marsh set the photograph down. "Because someone planned this, someone funded it, and someone lit the match."

The analyst hesitated just long enough to be noticed. "Third-tier contractor networks. The kind that don't appear in any formal tasking chain. Private, deniable, the sort of people who get encouraged rather than instructed."

Marsh understood immediately. Not a decision. An atmosphere. Someone had simply made clear that a spark would be welcome, and the spark had appeared.

"Encouraged by whom?"

"That," the analyst said, "is the question we're not going to be able to answer cleanly."

Marsh said nothing for a moment. "A spark designed to force a reaction."

"Yes, sir."

"Whose reaction?"

"Everyone's, ideally. That's the elegance of it."

***

Moscow

Volkov read the same report twice before he put it down.

"Not ours," he said.

His aide waited.

"I can see you want to ask whether it's useful to us." Volkov folded his hands on the desk. "It isn't."

"The narrative pressure on Ukraine—"

"Creates a moment, yes. Which then requires us to either exploit it or distance ourselves from it. Either choice tells them something about where we are." He shook his head. "The annex exists. The verification chain exists. If we escalate now, we are confirming that the annex threatens us. We hand them the frame they want."

The aide considered this. "So we say nothing."

"We say nothing, we move nothing, and we let them discover that the spark didn't catch." Volkov opened a different file. "Restraint is not weakness. Restraint is information management. Right now the most useful thing we can do is be boring."

***

Windhoek, State House

The emergency briefing had been going for eleven minutes when the President raised her hand and the room went quiet.

"Let me make sure I understand the situation. An oil facility has been struck, the narrative around the Maidan annex is being actively contested, and someone is waiting to see how we respond to both."

"That's accurate," Intelligence said.

"Is Namibia named anywhere?"

"Not directly. Not yet. But the framing of certain commentary is building toward questions about which non-aligned states have been, quote, cooperative with Ukrainian intelligence channels."

The President looked at Voss. "They want us to respond."

"They need us to respond," Voss said. "There's a difference. A response from us, any response, puts us inside the story. Right now we're outside it. That's exactly where we should stay."

Defence leaned forward. "No condemnation of the blast?"

"Not until attribution is clear and verified," the President said. "We don't condemn shadows."

"And if the attribution lands on Russia?"

"Then we note it through the appropriate multilateral channels, along with every other member state." The President's tone was measured. "It is not our war."

"And if it lands on the West?"

"Also not our war." She stood, and the room followed. "We do not react to sparks. We did not invite this fire, we will not help it spread, and we will not let anyone photograph us standing next to it."

***

The Manufactured Leak

A digital report surfaced simultaneously on three mid-tier investigative platforms, timed for the European morning news cycle.

Forensic Review Suggests Maidan Annex Digitally Altered Before UN Submission.

No smoking gun. No named source. Just technical language, checksum inconsistencies, timestamp anomalies, chain-of-custody questions that required specialist knowledge to evaluate and general anxiety to amplify. Plausible doubt, packaged for sharing.

Within hours, Western commentators were calling for independent verification. Russian outlets were mocking the original submission. Ukrainian social media fractured along lines that had been waiting to fracture.

The logic was clean: if truth cannot be destroyed, drown it in contestation until the audience gives up trying to find it.

***

Windhoek

Voss read the forensic challenge twice, then sat very still for a moment before she composed the encrypted message to Kyiv.

Integrity check required. Full chain verification. Assess whether public release of cryptographic proof is advisable.

Andriy's response came back in under four minutes.

Chain intact. Ethereum root hash unchanged since original time-stamping. We can publish the full cryptographic proof openly, every step, publicly reproducible. Anyone with basic technical literacy can verify it independently.

Voss read it again and didn't move.

If they released the proof, they walked directly into the confrontation. They were a small neutral state publicly demonstrating that a coordinated intelligence operation had failed. That had consequences.

If they stayed silent, the doubt would grow. Not because it was true, but because silence looked like it might be.

The President entered quietly and read the situation from Voss's expression before she read anything on the screen.

"Talk me through it."

"They've seeded a forensic challenge to the annex's authenticity. Technical language, no named source, designed to create enough noise that the underlying document gets buried in the dispute." Voss turned the screen slightly toward her. "We can prove the chain is intact. The cryptographic record is clean and independently verifiable."

"But?"

"But publishing the proof means we're in the fight. We're no longer a neutral conduit. We become a party to the confrontation."

The President was quiet. She walked to the window, then turned back. "They set off a small explosion. Then they seeded doubt about who built the device. Now they're waiting to see if the people holding the evidence go quiet or go loud." She looked at Voss. "What does going quiet tell them?"

"That we can be pressured."

"And what does it tell the Ukrainians? And the next person who needs a neutral channel to trust?"

Voss said nothing.

"We respond with facts," the President said. "Cold, technical, no commentary, no accusations. Just the verification record, published openly, and let anyone who wants to check it check it." She moved toward the door. "That is not escalation. That is what a country with nothing to hide does."

***

Global Feed

The cryptographic verification package was published through official Namibian government channels at 14:32 UTC. No press conference. No statement beyond a single line: *Full chain-of-custody verification available for independent review.*

Hash validation. Block-chain timestamp confirmation. Every step documented, every step reproducible by anyone with the technical means to check it. The kind of record that doesn't argue, it simply exists.

The forensic challenge collapsed within six hours. Western outlets that had amplified the anomaly questions quietly updated their coverage, replacing doubt with technical clarification. The blast in Venezuela lost its narrative oxygen as the story it was meant to ignite refused to catch.

The rally in Kyiv drew smaller numbers than the organisers had projected.

***

Langley

Marsh watched the verification cascade move through the monitoring feeds in silence. Around him the room had gone slightly still in the way rooms do when an outcome arrives that nobody wants to be the first to characterise.

"They didn't take the bait," someone said finally.

"No." Marsh kept his eyes on the screen. "They did something more inconvenient than that."

A pause. "Sir?"

"Taking the bait would have been useful to us. Going silent would have been useful to us. What they did instead was publish a standard." He turned from the screen. "Every neutral channel, every non-aligned state, every future actor considering whether to trust a verification process, they now have a reference point. A public, reproducible, technically rigorous reference point."

The room was quiet.

"Is that containment, sir?"

Marsh picked up his jacket. "No. It's evolution. And it's considerably harder to counter."

***

Windhoek

The intelligence report had been on the table between them for twenty minutes, and the President had not rushed through it.

When she finally set it down, Voss spoke.

"They will not forgive this."

"No," the President agreed.

"They came in expecting a small neutral country that could be nudged. What they found was more organised than that, and more principled than that, and now that's on record." Voss paused. "They'll recalibrate."

"And the internal pressure, the investor whispers, the stability questions?"

"We hold. We stay boring. The verification record helps. It's hard to run a political instability narrative about a government that just published impeccable cryptographic documentation."

The President nodded. Outside, the first pale light was beginning at the edge of the eastern horizon, pink deepening where the darkness had been.

"The next spark," Voss said quietly. "What does it look like?"

The President was still for a moment. "There won't be another spark." She looked toward the light. "Sparks are for testing thresholds. They've tested ours. They know what we are now."

"Then what comes next?"

"People," she said. "And violence. Not at us, necessarily. But close enough that we have to choose." She turned back from the window, and her voice was level. "So we should probably decide, right now, before it arrives, exactly what kind of country we intend to be when that moment comes."

Chapter 11: Quiet Removal
Kyiv

The radiator clicked in irregular intervals, metal expanding against winter pressure, a small domestic sound that should have been reassuring. Instead, in the borrowed apartment on Volodymyrska Street, it felt like a countdown.

Andriy Kovalenko had positioned the kitchen table at an angle so that he could see both the hallway and the reflection of the window in the dark screen of the television. Years of living under layered wars, military, digital, ideological, had taught him that threat rarely arrived from where you were looking directly.

The code on his laptop moved steadily, no longer urgent. The annex verification cascade had completed its last integrity sweep two hours earlier. Now it was merely replicating, threading itself into servers in jurisdictions whose political loyalties contradicted each other. Switzerland. Singapore. Reykjavík. São Paulo.

Mathematics, once released into distributed consensus, became stubborn. It resisted erasure.

Olena stood by the window without fully parting the curtain, her breath leaving a faint bloom against the cold glass.

Kyiv lay beneath a washed-out moon, the city quieter than it had been before the war, as though exhaustion had pressed sound into the ground. Somewhere far off, a tram bell rang once and faded.

"You notice it?" she asked. Her voice was low, but not tense.

Andriy did not answer immediately. He listened, not with his ears, but with the internal sense he had developed in the last week while monitoring global response.

The online landscape had flattened. No coordinated attack against the annex. No surge of nationalist fury or discrediting dossiers seeded into Western media. No disinformation spike from Moscow.

The ecosystem had gone still. "Yes," he said at last. "It's too quiet."

Silence in geopolitics did not mean indifference. It meant calculation.

He closed the laptop more carefully than necessary and slipped the secondary drive into his jacket pocket. He had rehearsed this motion enough times that it required no thought.

The hallway light outside flickered. Once. Then again, longer.

Olena's hand moved toward the pistol at the small of her back, not abruptly but with the calm inevitability of someone following a memorized sequence.

The building was old; power interruptions were common. But the flicker did not feel electrical.

Footsteps began climbing the stairwell. Not hurried. Not heavy. Measured. Even spacing between each landing. Professional pacing, conserving breath.

A knock came at the door. Soft. Two taps. "Maintenance," a male voice called in Ukrainian.

The accent was correct, but the cadence was not.

Olena's gaze met Andriy's for a fraction of a second.

No words were needed.

The first suppressed shot shattered the lock cleanly. There was no dramatic splintering, only a muted crack and the dull thud of the mechanism giving way.

The door opened inward with careful pressure.

Two men entered low, rifles compact, movements synchronized but not military-stiff.

Contractors often carried themselves differently from state soldiers; there was less ideological rigidity, more fluid adaptation.

Olena fired first. A single round, controlled. The nearest man folded backward, surprise registering only after his balance failed him.

The second attacker rolled toward the kitchen threshold, firing in short, disciplined bursts. Plaster erupted from the wall above Andriy's shoulder, filling the room with chalky dust. The refrigerator behind him ruptured with a hiss, a cloud of coolant spilling like breath.

Time contracted.

Andriy pivoted toward the rear room, not to flee blindly but to reach the fire escape, the exit he had measured three times since moving in.

The window was already opening inward.

A third figure climbed through, boots scraping brick, weapon rising with efficient certainty. They had mapped the apartment. This was not intimidation. This was removal.

For a split second, Andriy saw his own reflection in the dark window behind the intruder, pale, eyes wide, older than he remembered. He grabbed the nearest object, a metal desk lamp, and hurled it with more instinct than accuracy.

It struck the attacker's shoulder and fractured his aim. The shot tore through the ceiling instead of Andriy's chest.

Olena shifted position, using the refrigerator as partial cover, and fired twice. The man at the window collapsed awkwardly against the frame, one leg still outside, boot scraping brick in diminishing reflex.

The remaining attacker attempted to regain balance, but hesitation, that fractional human recalibration, was enough. Olena's third shot ended it.

Silence returned, but it was no longer neutral. It was heavy, occupied.

The radiator continued ticking.

Olena moved first, kicking rifles away, checking pulses with swift efficiency. The smell of gunpowder mixed with leaking coolant and old cabbage from a neighbouring flat.

"No insignia," she said quietly.

Andriy knelt beside the nearest body. The man's gloves were tactical but unmarked. His face could have belonged to Warsaw, Vienna, or Marseille. There were no ideological tattoos, no nationalist symbols.

Only a compact encrypted handset clipped to his vest. Custom hardware. Baltic components.

Not state-issued.

Which meant plausible deniability. Andriy felt something colder than fear move through him.

State actions could be predicted, but private networks were less disciplined. Less accountable and a lot more dangerous.

Olena straightened slowly. "They knew the layout."

"Yes."

"They weren't improvising."

"No."

He looked at the dead man's face again and felt no hatred, only recognition.

The annex had ceased being an argument between governments. It had become a variable in someone's financial model.

And that, he suspected, was more dangerous.

***

Windhoek

Rain had washed the city clean overnight. The jacarandas along Independence Avenue glistened faintly in the early light, their branches heavy, the dust pressed flat against pavement and memory alike.

President Nadi-Omunongo stood alone in her office before the staff arrived, a mug of tea cooling untouched on the low table behind her. The thunder had kept her awake for hours. Storms always unsettled her, not because of fear, but because they reminded her of how small structures were against accumulated force.

The intelligence report lay open on her desk.

Kyiv. Attempted removal. No official attribution.

She read the lines twice, then closed the folder and walked toward the window.

At seventy-three, she had learned to distrust moments when events accelerated too neatly. The oil discoveries offshore had already drawn interest sharper than diplomacy.

Total commitments. European delegations. American hypocrisy masked as partnership. Russian curiosity disguised as cultural exchange.

Now, private actors were moving in Eastern Europe to control information that might shift energy narratives.

Her reflection in the glass showed a woman who had survived more than she had ever intended to. She had entered politics believing in sovereignty as something almost sacred.

Over the decades, she had watched sovereignty diluted by loans, by aid packages, by "technical assistance," by joint ventures whose contracts were thicker than constitutions.

She allowed herself a rare private thought:

Perhaps neutrality was no longer protection.

Perhaps it was bait.

If Namibia was about to step into the global energy arena, it would not be permitted innocence, and for the first time since the oil announcement, she felt not ambition, but apprehension.

Not for herself, but for her country.

She straightened as footsteps approached in the corridor. The vulnerability retreated behind practiced composure.

When the door opened, the President of Namibia was once again entirely controlled.

Chapter 12: Orion
London

The building had been commissioned in 1894, when the empire still painted maps in red and called extraction "civilization."

Orion Global Energy had never moved.

Its headquarters stood on a discreet street off St. James's, stone façade soot-darkened, columns fluted in an era when oil was still whale fat and ambition wore a waistcoat. The brass plaque by the door had been polished so often that the letters seemed engraved by repetition rather than design.

Inside, the lobby smelled faintly of old paper and polished wood.

There were no digital welcome screens. No sustainability slogans. Only portraits.

Men in tailored suits, severe expressions, one boot resting on crates marked with colonial port names, Lagos, Mombasa, Basra. The frames were heavy gilt. The captions understated.

Founding Director.

Regional Administrator.

Expansion Committee Chair.

No mention of what had been expanded.

The boardroom occupied the top floor, beneath a coffered ceiling darkened by a century of cigar smoke. The windows were tall but narrow; light entered reluctantly, as if aware of its function here.

Sir Malcolm Havers sat at the head of the table. His family had held Orion shares before most nations had held elections.

His grandfather had overseen pipeline concessions in West Africa during the years politely referred to as "transitional governance." His father had sat on three advisory boards whose minutes were never published. Sir Malcolm had learned early that the quietest rooms made the loudest decisions.

He disliked modern language, ESG metrics, impact audits, stakeholder dialogue. To him, those were cosmetic burdens imposed by younger executives who had never negotiated with generals in desert airfields. Sentiment, he believed, was inefficiency wearing a conscience.

He folded his reading glasses and placed them on a leather blotter. "Namibia," he said, not loudly, but with the tone of someone placing a chess piece.

Around the table, directors shifted papers, not because they needed them, but because ritual demanded it. Sir Malcolm had always appreciated that about his board. They understood performance.

A map illuminated on the far wall. Offshore blocks. Exploration grids. Seismic overlays.

"New basin," said Adrian Cole, Orion's Director of Strategic Expansion. "High-grade light crude. Politically stable jurisdiction. Under-capitalized infrastructure. Limited domestic expertise in deep-water extraction."

Translation: accessible.

Sir Malcolm leaned back slowly. "And the Americans?"

"Watching."

"And Moscow?"

"Circling."

"And Beijing?"

"Present, but cautious."

Sir Malcolm allowed a thin smile. "Good. Crowded rooms create urgency."

There was no mention, not yet, of the Kyiv incident. Not yet of the annex verification that threatened to disrupt European energy narratives. But it was understood in the way seasoned predators understand shifting winds.

Energy markets were psychological before they were geological. A younger director, freshly installed after a compliance scandal in Southeast Asia, cleared her throat. "We should consider reputational exposure. Namibia has a strong civil society network. The President—"

Sir Malcolm's gaze moved to her, not hostile, but corrective.

"President Nadi-Omunongo is seventy-three. Stability at that age is transitional by definition."

The room absorbed that quietly.

She did not speak again.

Adrian Cole intervened smoothly. "We have already activated philanthropic channels. Educational grants. Coastal infrastructure partnerships. Health initiatives."

Sir Malcolm nodded once.

Carrots first.

"And the less visible channels?" he asked.

There was no discomfort in the room. Only professionalism.

Cole adjusted his cufflinks. "We've engaged Orion Advisory Services to map political fracture points. Youth unemployment rates are high in Katutura and Walvis Bay. Digital influence campaigns can amplify governance dissatisfaction without direct attribution."

"External actors?"

"Subcontracted. Eastern European networks. Plausibly deniable."

Sir Malcolm tapped the table lightly with one finger.

"And local catalysts?"

"Two opposition figures are receptive. One believes this is about transparency. The other believes it's about succession."

Sir Malcolm gave a small, satisfied exhale.

Idealism and ambition. The two most renewable fuels in politics.

"And Kyiv?" he asked finally.

Cole hesitated only slightly. "Handled."

"How?"

"Through an intermediary security consortium. The objective was disruption, not spectacle."

Sir Malcolm's expression did not change. "Disruption must be definitive, Adrian."

A pause. "We are not in the business of half-measures."

Outside, London traffic moved obediently beneath a grey sky. Inside, the air felt older than weather.

A portrait hung behind Sir Malcolm's chair, his great-grandfather standing beside a drilling rig in what had once been Northern Rhodesia. A local labour foreman stood three paces behind him in the photograph, unnamed.

The empire had not vanished. It had incorporated.

"Proceed with Phase Two," Sir Malcolm said quietly.

"Civil unrest simulation?" asked Cole.

"Correction," Sir Malcolm replied. "Civil unrest encouragement."

The younger director shifted again. "And if Moscow positions itself as benefactor to Namibia?"

Sir Malcolm smiled faintly. "Russia does not invest without expectation. If they step in publicly, Western capital will respond reflexively. Markets dislike ideological romance."

He rose, signalling the meeting's end.

As the directors gathered their documents, Sir Malcolm remained standing, looking out the narrow window toward the pale Thames.

He did not think of Namibia as a nation. He thought of it as a variable. Reserves quantified. Timelines projected. Governance manageable.

***

Orion Advisory Services
Riga, Secure Office

The activation order arrived at 14:23 Central European Time. No letterhead. No names. A single encrypted line routed through three proxy relays: Phase Two. Windhoek. Begin.

The three analysts in the room did not react visibly. This was not their first basin.

Within the hour, dormant social media accounts in Namibia began producing content. Slowly at first, calibrated below anomaly thresholds. Youth unemployment figures, reshared without attribution. A video clip of a cabinet minister's motorcade, stripped of context, framed as excess. A question, planted in a comment thread beneath a local news article: Who benefits from the oil when ordinary Namibians cannot afford bread?

The accounts had no connection to each other. No shared IP signature. No pattern a basic filter would catch.

Engagement metrics climbed at a controlled rate.

"Velocity at 2.1 percent," one analyst said, watching the dashboard.

"Hold there," replied another. "Let it breathe."

The third said nothing. He was already seeding the next thread.

Outside, Riga's old city caught the grey afternoon light. Inside, the work continued with the quiet efficiency of people who had learned long ago that the most effective pressure leaves no fingerprints.

***

London

Sir Malcolm remained at the window for a long moment after the room emptied.

What he did not yet account for, what Orion had historically underestimated, was what happened when smaller states refused to fracture on schedule. History offered occasional, inconvenient examples. He had always attributed those to poor execution.

He did not consider, not yet, that the variable might be something other than execution. That it might be will.

Chapter 13: Fracture Lines
Windhoek

The rain had stopped, but the streets still held its memory.

Puddles collected along the cracked pavements of Katutura, reflecting broken telephone wires and satellite dishes that leaned at uncertain angles. The air smelled of wet dust and diesel.

Lukas Shilongo stood outside the small internet café on Eveline Street, his phone vibrating in his hand for the third time in ten minutes.

The video had appeared just after lunch, of a grainy clip of offshore drilling projections. A caption in bold red:

WHO PROFITS FROM YOUR OIL?

Below it, a shorter line: NOT YOU.

The account that posted it had no history. No location. No followers before this morning. By mid-afternoon it had thousands.

Lukas wasn't a revolutionary. He was twenty-four, unemployed despite a technical diploma, living with his aunt, and angry in the unfocused way that comes from stalled momentum.

He tapped the screen again. Another post:

Foreign deals signed behind closed doors.

Elites grow rich, as youth remain jobless.

There were screenshots, contracts, real or convincingly fabricated, highlighting clauses about profit-sharing structures and foreign technical partnerships.

In the comments, something had shifted. At first, frustration. Then certainty. Then direction.

"Protest tomorrow," someone had written. "Independence Avenue. 10:00."

Another account amplified it. Then another.

Lukas felt the subtle pull of alignment, that dangerous human sensation of being part of something larger.

His aunt called from inside. “You coming in or just staring at that thing?”

He didn’t answer, as he pressed “share.”

***

Orion Advisory Services
Riga, Secure Office

The room was windowless and aggressively modern, glass desks, filtered air, and ergonomic precision. No portraits. No history.

Three analysts monitored dashboards where engagement metrics rose in controlled increments.

“Windhoek trending at 4.2% velocity,” one of them said.

“Keep amplification below anomaly threshold,” replied another. “We want grievance, not algorithmic suspicion.”

Payment confirmations flashed through layered accounts, Baltic shells, consultancy retainers, subcontracted creative firms in Bucharest.

Orion’s name did not appear anywhere. It did not need to.

***

Windhoek

The sky was pale and undecided. By nine-thirty, groups had begun to gather along Independence Avenue. Not thousands, not yet. But enough to change the sound of the city.

Handwritten signs appeared first. Then printed placards that looked suspiciously professional for overnight work.

TRANSPARENCY NOW

OUR OIL, OUR FUTURE
STOP SECRET DEALS

Lukas found himself near the front without meaning to. Someone handed him a sign. He took it.

A small drone hovered above, live streaming.

At first, it felt almost festive, a civic release valve opening.

But at the edges of the crowd, men stood who did not chant.

They observed.

And when the first glass bottle flew, no one could later say from where, it shifted the temperature instantly.

Police lines tightened. Shields rose.

The chant fractured into shouts, as a second object struck a patrol vehicle. Then a third.

Someone screamed as the sound of a stun grenade detonated against wet pavement echoed down the avenue, violent and concussive. Lukas felt the shock in his chest before his ears registered it.

Panic moved faster than ideology, as the crowd surged backward, trampling its own edges.

Smoke drifted low across the street, and through it, a handful of masked figures, not local, not familiar, moved with purpose, hurling heavier objects toward government buildings.

Escalation inserted.

***

State House, Windhoek

President Nadi-Omunongo watched the live feed without speaking.

The rain-washed city she had looked upon yesterday now shimmered through tear gas haze.

"Spontaneous?" she asked quietly.

Her intelligence chief did not answer immediately. “No, Madam President.”

She knew that already.

The digital forensics team had flagged coordinated account creation patterns overnight, time stamps inconsistent with Namibian usage rhythms. Payment routes traced to European intermediaries.

But that was only one vector.

She turned away from the screen and walked toward a side table where a different briefing lay open.

Chinese Infrastructure Holdings, Namibia Portfolio Overview.

Uranium mining stakes in the Erongo region. Walvis Bay port modernization, majority operational control through long-term concession. Highway expansions connecting inland corridors to maritime export routes. Billions invested.

The Chinese ambassador had called at dawn. His tone had been measured but courteous. “China values stability in Namibia,” he had said.

She heard the unspoken continuation clearly: *instability would be corrected*. She understood great powers. She had negotiated with enough of them.

The Americans would issue statements about democratic process.

The Europeans would express concern.

Moscow would position itself as respectful of sovereignty.

Beijing would not posture. Beijing would protect infrastructure, and infrastructure in Namibia increasingly bore Chinese signatures.

She felt, briefly, the weight of geography. Small states did not choose storms. They endured them.

***

Langley

Elliot Marsh replayed the Windhoek footage alone. He had requested it without commentary. No analyst summaries. No strategic overlays. Just raw feed.

He watched the moment the bottle flew. Paused it. Rewound.

The masked men at the edge moved too deliberately. The spacing between them suggested training. "External agitation," he murmured.

A younger analyst entered quietly. "Sir, Chinese communications intercepts indicate heightened alert posture regarding Walvis Bay."

"Military?"

"Not overt. But security contractors repositioning."

Marsh exhaled slowly. "So we have Orion destabilizing. Moscow positioning. Beijing hardening assets."

"And us?" the analyst asked.

Marsh didn't answer immediately. For decades, American intelligence had assumed that global pressure points would eventually resolve into a binary contest, with the US always as the victor.

Now, it felt multipolar, and privately financed.

"Monitor," he said finally. "Do not engage yet."

The analyst hesitated. "Sir, if we don't shape this?"

Marsh cut him off, but gently. "Sometimes shaping becomes ownership. And ownership becomes liability."

When the analyst left, Marsh remained seated in the darkened office. He had joined the Agency believing that stability, however imperfect, prevented catastrophe.

Now he was watching private corporations ignite unrest in a small African democracy to secure offshore drilling rights, while China quietly prepared to secure its uranium and port investments by any means necessary.

And somewhere in Moscow, he suspected, Volkov was calculating how to turn Western overreach into strategic advantage.

Marsh rubbed his eyes. The world was no longer just about ideology. Ideology has been replaced by shadow organizations manipulating governments to secure oil, mineral resources, and infrastructure for private profit. And resources, once threatened, triggered responses that were rarely subtle or diplomatic.

Chapter 14: Pressure Points

Windhoek

The rally had dispersed hours earlier, but the city had not returned to normal.

The air still held a faint chemical sharpness from tear gas. Along Independence Avenue, broken glass glittered beneath streetlamps like scattered frost. Police trucks idled at intersections long after the crowd had gone.

President Nadi-Omunongo insisted on driving home.

Not in the armoured convoy, though it shadowed discreetly, but in the official sedan she had used for years. She believed routine was stabilizing. A nation watched for signs of fear.

Inside the vehicle, the rain had returned in a fine mist, streaking the windows in diagonal lines.

Her driver adjusted his grip on the wheel as they turned toward the quieter residential stretch near Eros.

The first motorcycle appeared in the side mirror almost invisibly. No headlights. Just a dark outline moving in parallel.

The driver noticed it. He did not react immediately.

Protocol dictated confirmation before escalation.

The motorcycle accelerated suddenly, closing distance with predatory smoothness. A second bike emerged from a side street ahead.

Too precise.

The driver's pulse spiked. "Madam President," he began.

The first shot struck the rear quarter panel with a violent metallic crack.

Not a warning.

The second shattered the rear window.

Glass exploded inward in a crystalline burst. The President instinctively lowered herself, the movement not rehearsed but reflexive, the body remembering survival even when the mind resisted vulnerability.

The driver swerved sharply, tires screaming against wet tar. The escort vehicle behind surged forward, headlights flaring.

Gunfire multiplied, controlled, not chaotic. This was not rage. It was execution.

One motorcycle veered closer, rider lifting his arm for clearer angle. The escort vehicle rammed it deliberately.

Metal collided with bone and machine in a sickening cascade. The rider flew sideways, skidding across rain-slick asphalt.

The second motorcycle aborted, disappearing into a narrow side street before perimeter units could seal it.

Silence followed, broken only by the ticking of the engine and the President's steady breathing.

She raised herself slowly. Blood from a shallow cut along her temple traced a thin line toward her collar.

"I am fine," she said before anyone asked.

Her voice did not tremble. But inside, something colder had taken hold.

This had crossed a threshold. Unrest was influence. This was an attempt on her life, and she do not take kindly to it.

***

Kyiv, Undisclosed Location

Andriy adjusted the compression bandage on Olena's shoulder with more care than she would have used on herself.

"It grazed," she insisted.

"Yes."

The safe house this time was not an apartment but a basement beneath a print shop whose machines had not operated in months. Ink and dust mingled in the air.

The Kyiv attack had confirmed what Andriy already suspected: the annex threatened interests beyond governments.

His phone vibrated once, encrypted channel.

Daniela Ruiz.

He hesitated before answering.

Her face appeared on screen, dimly lit by what looked like a hotel room in Brussels.

“You’re alive,” she said.

“Temporarily.”

She didn’t smile.

“I’ve traced the Baltic intermediaries from the attempted hit. The shell structures intersect with a consultancy tied to Orion’s litigation defence network.”

Andriy felt the pattern locking into place. “They’re protecting energy exposure,” he said.

“Yes. But not only that.”

She leaned closer to the camera.

“Chinese state banks are accelerating loan guarantees for Namibian infrastructure this week. It’s defensive positioning. They anticipate instability.”

“And Russia?”

“Quiet. Which means waiting.”

Olena listened without interrupting.

Daniela exhaled.

“Andriy… this is no longer about annex verification. It’s about who controls the next energy corridor.”

He knew, and that was why he was still alive. For now.

***

Somewhere Outside Vienna, Secure Line

The meeting between Volkov and Marsh did not occur in person.

It occurred through silence first. Then through a single indirect message routed through an academic energy symposium scheduled in Istanbul.

A footnote in the agenda referenced a *"closed technical discussion"* on sovereign infrastructure resilience.

Marsh recognized the phrasing immediately. He approved attendance. Volkov would not threaten. He would offer.

***

Windhoek

Chinese Air Force transport aircraft rarely landed in Namibia.

But this one did, officially carrying a "technical advisory delegation" to review port security cooperation at Walvis Bay.

The arrival was courteous and professional, with engineers in tailored suits. Security advisors whose posture betrayed prior military service.

The Chinese ambassador stood on the tarmac beside Namibian officials, smiling calmly as cameras flashed.

"China stands with Namibia in ensuring stability and mutual prosperity," he said. No mention of unrest. No mention of the assassination attempt.

But the delegation did not travel only to Walvis Bay. Two vehicles quietly diverted inland toward Windhoek that afternoon.

Infrastructure required assessment. So did political risk.

***

Namibia State House

President Nadi-Omunongo sat alone before meeting the Chinese delegation.

The bandage had been replaced discreetly. Makeup concealed what it could.

She had survived an attempt on her life. And yet what troubled her more was the implication: someone believed Namibia small enough to destabilize without consequence.

She no longer believed that. Outside her office window, the city moved cautiously.

She thought of uranium in the Erongo desert. Of deep-water oil beneath Atlantic currents. Of ports. Of roads stretching inland. Namibia was no longer peripheral, it has become an uncontrolled intersection. And intersections attracted collision.

She straightened as her aide entered.

"The Chinese delegation is ready, Madam President."

"Let them in." Her voice was steady.

But in the quiet space behind it, apprehension had begun to replace calculation.

Chapter 15: River of Shadows
Colombian, Venezuelan Border

The jungle did not allow speed without punishment, as the van bucked violently as it tore along the mud-carved track, suspension shrieking with each rut. Weeks of rain had liquefied the earth; what had once been a service road was now a corridor of brown slurry bordered by dense, breathing green.

The Orinoco moved parallel to them on the left, swollen beyond its banks, the current twisting and folding over itself with muscular force.

Daniela Ruiz braced her uninjured arm against the dashboard as another jolt sent pain lancing through her shoulder. The bullet graze had stiffened overnight. Dried blood pulled against fabric each time she moved. The smell inside the van was metallic, sweat, river air, fear.

Carlos drove without speaking, both hands fused to the steering wheel as if it were the only stable object left in the world. The scar along his cheek, usually hidden beneath stubble and humour, stood pale against his damp skin.

"Border's close," he said finally, voice tight but controlled. "Your family crossed at dawn. Colombian side."

Family. The word did not comfort her. It hollowed her.

She had watched the leak detonate from three different devices in three different safe locations. What had begun as documentation, contracts, payment chains, advisory structures linking energy conglomerates to destabilization networks, had turned into global combustion within hours.

Oil futures spiked.
Television panels filled.
Governments denied.

Markets convulsed faster than armies ever had.

She had wanted exposure, but did not fully calculated velocity.

Gunfire cracked behind them. Not random. Measured bursts.

Carlos did not look in the mirror immediately. He corrected the steering first, instinct before confirmation.

Then he glanced back. Two pickup trucks had forced their way through the brush behind them, tires spitting mud. Flags whipped violently from antenna poles, 'colectivos' insignia, and red against green jungle.

Masked men stood in the truck beds. Rifles already shouldered. The first burst shattered the van's rear window, and glass exploded inward, scattering across Daniela's lap like ice. A small shard lodged in her cheek. She did not feel it yet.

Carlos swerved left to avoid a fallen branch; the tires lost purchase entirely. The van fishtailed, corrected, and then clipped a tree with a blunt crack that reverberated through Daniela's ribs.

Another volley. Closer.

Then Carlos made a sound she had never heard from him before, short, surprised.

She turned.

His chest seemed to cave inward, as if the air had been punched from him permanently. Dark blood spread rapidly across his shirt, soaking into fabric that moments earlier had been ordinary.

His hands loosened and the steering wheel spun free. The van left the track entirely and dropped nose-first into a shallow ditch. Metal screamed. Steam hissed from beneath the crumpled hood. Daniela's forehead struck the side pillar hard enough to blur her vision white.

For several seconds there was only the roar of the river.

Carlos slumped forward, eyes open but already unanchored.

She stared at him.

There are moments when grief attempts to enter. Survival blocked it. She forced the passenger door open with her shoulder and stumbled into knee-deep mud.

The backpack was still wedged between her feet. She grabbed it automatically. Inside it: external drives. Redundant encryption keys. Contact lists she no longer trusted.

It felt heavier than electronics should.

She leaned back into the van once more and fumbled through Carlos's jacket, retrieving his phone. The screen lit weakly. A message. Two minutes old.

Secure asset before Colombian transfer. Window closing.

No signature. No identifier. No ideology. Just instruction.

Asset. The word reconfigured something inside her. She was no longer the author of events. She was inventory.

Boots crashed through foliage behind her. Voices shouted, sharp Venezuelan Spanish, edged with fury.

She ran. The jungle closed instantly around her. Vines caught at her ankles. Branches tore at her face. Humidity pressed against her lungs like wet cloth. Every step threatened to twist her injured leg.

Bullets tore through leaves overhead, shredding canopy into green rain.

She did not run toward safety. She ran towards the river. The tributary appeared abruptly, swollen and violent. Brown water frothing over hidden rocks. There was no bridge, only a scatter of slick stones barely visible beneath the surface.

Behind her, the 'colectivos' reached the bank first, firing wildly into the waterline.

Then, from the opposite flank, figures emerged from the undergrowth. Not colectivos. Different discipline.

Their movements were tighter, more economical. Weapons held with practiced economy rather than performativity aggression. Faces streaked with camouflage paint.

One of them spoke into a radio.

"Target visual. Persian window compromised."

Persian?

The word lodged in her mind even as her body reacted.

Another layer. Another interest. She did not have space to decode it.

She stepped onto the first rock. It betrayed her immediately, moss slick as oil. She wind milled for balance, boots skidding. The current slammed into her calves with brutal force, trying to fold her sideways.

A round struck a stone inches from her hand, ricocheting with a metallic scream.

The backpack dragged at her spine like an anchor. Water surged higher, slamming against her thigh wound. Pain flared hot and disorienting.

A bullet tore into the water beside her leg, close enough that she felt its passage as heat before cold swallowed it.

She slipped.

Her knee smashed against submerged rock and she pitched forward into the current. Brown turbulence engulfed her. For one disorienting breath, there was no sky, only sediment and force.

The river did not care who funded what. It only pulled.

She clawed blindly, fingers scraping algae and stone, lungs burning. Somewhere above the roar, gunfire continued, but farther now, uncertain.

She surfaced, gasped, and dragged herself toward a final jut of rock near the Colombian bank.

“Let the river take her!” someone shouted from the Venezuelan side, but no one followed.

Borders still mattered.

She lunged forward, fingers sinking into mud and reeds. The current fought to reclaim her, but she dug deeper, hauling herself onto the bank in a graceless collapse. Her body lay half in water, half in soil. Alive.

She did not remember standing. Only stumbling inland, water pouring from her clothes, blood mixing with mud in dark streaks along her leg. The jungle thinned gradually into scattered huts and cooking smoke.

A Colombian village materialized like a fragile mirage, wooden structures leaning against heat and poverty in equal measure. Children stopped mid-play to stare. A dog barked once and retreated beneath a cart.

Daniela leaned against the outer wall of a hut and slid downward until she was seated in dust.

A lean man approached cautiously, machete hanging at his hip. His eyes were not unkind, only calibrated.

“You crossed from trouble,” he said. She nodded.

“Trouble crosses back.”

“I need time,” she whispered.

“Time costs.”

She lifted the backpack weakly.

“What’s inside?” he asked.

“Information.”

He studied her longer than comfort allowed. "In this region," he said finally, "information buys survival only if someone wants it."

"Someone does." He considered that. Then he extended his arm and helped her stand.

Inside the hut, darkness folded around her like thick cloth. The door was barred. Outside, distant helicopter rotors began to thud against the canopy, heavy, searching.

She pressed the backpack to her chest. Asset. Window closing. Persian?

The leak had not merely exposed Orion. It had intersected with something larger, energy corridors, sanctions structures, shadow financing, and state interests that did not publish press releases.

She had believed exposure would weaken empires.

Now she understood the inversion. Exposure revealed value.

And value invited acquisition.

Somewhere far from the river, men in quiet rooms were recalculating her worth. And several of them, she suspected, had already decided she was expendable.

Chapter 16: Fractures Deepen

Moscow

Dmitri Volkov stepped out of Lubyanka's side entrance, collar up against the biting wind, snow crunching under his boots. The leak was live, the world twisting as planned, CIA exposed, Iran howling accusations of "oil wars" as prelude to U.S. attacks. Geneva "peace" talks crumbling in finger-pointing,

US President Harvey tweeting threats of "consequences" for "leakers and liars." But Volkov knew the real game: FSB's hardliners saw his subtle plays as weakness, exposure as surrender. They wanted war and full escalation in Ukraine, no "soft" leaks to weaken the West over time.

He slid into his unmarked Lada, engine turning over with a growl, heading for a dead drop in the Arbat district. The streets were slick with slush, traffic a slow crawl through the snowstorm. In the rear-view, a black SUV edged closer, too close. Volkov's instincts flared.

He accelerated, weaving through cars, but the SUV matched, ramming his bumper with a thud that spun him toward a snowbank.

Ambush. Hardliners, his own side. Volkov floored it, tires spinning on ice, the Lada fishtailing into a side street. The SUV pursued, windows rolling down, muzzles flashing. Bullets pinged off his trunk, shattering the rear glass in a shower of shards.

He yanked the wheel, cutting through a market alley, with vendors scattering, stalls toppling in his wake. Snow blinded, but he knew Moscow's veins.

The chase intensified: Volkov dodged into an underground parking garage, ramps spiralling down, and tires screeching echoes off concrete. The SUV followed, closer now, a second vehicle joining from a ramp, reinforcements.

Gunfire erupted in the confined space, bullets ricocheting like hornets. Volkov braked hard, reversing into the lead SUV with a crunch of metal, then leaped out, pistol drawn.

He rolled behind a pillar as rounds chipped concrete. Two assassins advanced, hard-line FSB, faces masked, suppressed rifles barking. Volkov fired back, dropping one with a headshot, blood spraying the wall.

The second closed in, knife drawn for silent work. The blade slashing across his ribs, Volkov ignore the sharp pain and deflect the blade, elbow smashing the man's nose, bone crunching.

They grappled, slamming into cars, alarms blaring chaos. Volkov twisted the knife free, plunging it into the assassin's inner-thigh femoral artery, then a chokehold until the body went limp.

Breathing ragged, wounds shallow but bleeding, Volkov staggered to a stolen sedan in the shadows, always a backup.

He peeled out, the garage behind him a tomb. The FSB was split: Some, like him, saw value in exposing the CIA's long game to weaken the West. Others craved war's fire. He was hunted now, by his own. But the plan endured for now, as he has a meeting to attend that might bring more clarity.

***

Istanbul

The Bosphorus at night did not belong to Europe or Asia.

Ships moved slowly through the black water, their lights smeared by mist and distance. Ferries crossed in deliberate arcs. The city rose in layers behind them, minarets, glass towers, satellite dishes, and the faint hum of ten million private calculations.

The energy symposium had concluded an hour earlier.

Panels had discussed “resilience frameworks” and “multipolar infrastructure security” in language that neutralized meaning through repetition. Delegates had shaken hands beneath banners printed in blue gradients.

Now, in a private room above a restaurant overlooking the strait, two men sat across from one another without aides.

Elliot Marsh arrived first. He chose the chair with his back to the wall and a partial view of the door. Old habit. He ordered tea instead of wine. His reflection in the dark window looked more tired than he remembered.

When Volkov entered, he did so without ceremony. No entourage. No visible security.

Just a dark coat and the unhurried posture of a man who had learned long ago that appearing unthreatened was often more destabilizing than appearing armed.

They shook hands.

Not warmly. Not coldly. Professionally.

Volkov removed his coat and sat.

For a moment they did not speak. The restaurant below hummed faintly, cutlery, muted conversation, the soft friction of commerce.

“You came,” Volkov said at last.

“You invited,” Marsh replied.

A waiter entered, placed a small glass of tea before Volkov, and left without acknowledging either man directly.

Silence returned.

Marsh broke it. “Namibia.”

Volkov’s expression did not change. “Yes.”

“You’re positioning.”

“We are observing.”

Marsh gave a faint, humourless smile. “You’re never only observing.”

Volkov accepted that without denial. “The West destabilizes when it overreaches,” he said quietly. “It is a recurring pattern.”

Marsh leaned back slightly. “Orion is not the United States.”

“No,” Volkov agreed. “But Orion believes it can move where governments hesitate.”

The implication lingered between them, as private capital had crossed a line in Kyiv, then Venezuela, and now looking at Namibia.

Marsh did not confirm that he knew.

Volkov did not confirm that he had intelligence beyond what was public.

“You intend to offer Namibia protection,” Marsh said.

Volkov stirred his tea slowly. “Protection is a Western word,” he said. “We offer partnership.”

“And what does partnership cost?”

Volkov met his eyes fully for the first time. “Alignment.”

Outside, a tanker moved through the Bosporus, engines low and constant.

Marsh considered his next words carefully. “China is already embedded,” he said. “Ports. Uranium. Roads. They will not tolerate instability.”

Volkov nodded. “China protects investments and infrastructure. We protect relationships.”

“And Orion?”

Volkov allowed the smallest shift in expression, not quite amusement. “Orion protects extraction and profits.”

The distinctions were deliberate.

Marsh felt something tighten beneath his sternum.

For decades, Washington had assumed that its adversaries would operate through states, not corporations. That conflict would be ideological, not financial. That destabilization would carry flags.

Now the flags were gone.

"Did you intervene in Kyiv?" Marsh asked quietly.

Volkov did not answer immediately.

"The annex," he said instead, "threatens multiple interests. Some prefer it unresolved."

"That wasn't my question."

"No," Volkov agreed. "It was not."

The air between them cooled slightly.

Marsh studied him. "You're letting the West damage itself," he said.

Volkov's response was calm. "The West is very capable of that without assistance."

A faint call to prayer drifted across the water from a distant mosque, folding into the night.

Marsh shifted the conversation. "If Namibia fractures," he said, "it becomes a corridor. Oil, uranium, Atlantic access."

"Yes."

"And if private actors push too hard?"

"Then smaller states look for counterweight." Volkov answered.

Marsh understood the offer now.

Russia would not attempt overt dominance. It would position itself as sovereign-friendly. Respectful.

Reliable in contrast to Western corporate coercion.

"And what do you want from us?" Marsh asked.

Volkov did not hesitate. "Restraint."

Marsh almost laughed. "That's a luxury."

"It is a strategy," Volkov replied.

Another silence.

Marsh thought of Windhoek. Of the assassination attempt. Of Chinese aircraft landing under infrastructure pretext. Of Daniela Ruiz bleeding into jungle mud. Of Andriy nearly eliminated by contractors whose payroll seem to trace back to energy litigation structures?

The world was no longer divided cleanly between adversaries. It was divided between those who could absorb shock and those who could not.

"Namibia survives if it is not forced into binary choice," Volkov said quietly.

"And if it chooses?" Marsh asked.

"Then we adapt, as Russians always do."

The simplicity of it unsettled Marsh more than aggression would have.

Volkov finished his tea and stood. "We are not enemies by necessity," he said.

"That depends on definitions." Marsh say softly.

"Yes," Volkov agreed. "It always does."

They shook hands again.

As Volkov left, Marsh remained seated, watching the dark water move through the strait.

Restraint. It sounded almost noble.

But he knew Washington would not remain still if it believed Moscow was gaining foothold in Atlantic energy corridors. And Beijing would not remain passive if port security was threatened. Orion would not relinquish an oil basin without leverage or a fight, there is too much money involved.

The new Global multi-polarity has not brought equilibrium, only heighten tension.

The question was no longer who would dominate Namibia.

It was who would miscalculate first.

Chapter 17: Convergence
Walvis Bay, Atlantic Ocean, Namibia

Morning light broke through low marine fog, turning the cranes of Walvis Bay into skeletal silhouettes against the pale horizon. Container ships sat in disciplined rows offshore, their hulls painted in corporate blues and reds, each one a floating artery in a global circulatory system that never truly slept.

The Chinese delegation arrived at the port not as spectacle, but as procedure. Clipboards. Tablets. Quiet cameras.

Rear Admiral Liu Wen walked the quay beside Namibian Port Authority officials, his expression measured, neither smiling nor stern. His civilian attire concealed nothing about his background; posture alone suggested decades of disciplined command.

"Security perimeters appear adequate," Liu observed, pausing near a newly installed scanning system manufactured in Shenzhen. "However, layered instability inland can migrate toward maritime chokepoints."

The Namibian port director stiffened slightly. "We do not anticipate migration."

Liu nodded politely. "Anticipation is most effective before confirmation."

Behind them, two members of the delegation drifted subtly away from the main group. They did not photograph cranes or cargo. They photographed sight lines. Road access points. Fuel storage positioning. Police station proximity.

China had invested billions in this coastline, in uranium extracted from Erongo's pale desert, in highways cutting through dry riverbeds, in the deepening of this harbour.

Investment did not tolerate volatility.

By afternoon, an unpublicized meeting was scheduled in Windhoek between the delegation and select officers of the Namibian Defence Force.

Not as interference. As "technical consultation."

***

Windhoek, Military Headquarters

Colonel Petrus Hamutenya had served under three presidents and through two constitutional reforms. He believed in hierarchy, in process, in the slow grind of institutional continuity.

He did not believe in foreign advisory presence inside operational rooms.

Yet he now sat across from Liu Wen in a quiet conference chamber, listening as the Chinese officer spoke in unhurried tones about "infrastructure protection doctrine."

"We have observed," Liu said, "that private actors often introduce volatility to secure negotiation advantage. It is important for sovereign states to maintain clarity of force posture."

Hamutenya folded his hands. "Namibia is sovereign," he said evenly.

"Of course," Liu replied. "We are here because you are."

The distinction was razor-thin.

Later that evening, Hamutenya would receive a separate call, routed through a secure channel not officially linked to the Chinese delegation, offering advanced crowd-control equipment and cyber-monitoring tools at concessionary rates.

Stability, subsidized.

***

Windhoek, Internal Security Directorate

Inside a smaller, less ceremonious building, another fracture had begun to widen.

Deputy Director Elias Kandjii stared at the protest footage again, isolating faces frame by frame.

He had already identified three individuals whose biometric signatures did not match Namibian records. Baltic origin probable. Contractors.

He had also identified something else.

An internal clearance code had been used to delay riot deployment by nine minutes at the height of escalation.

Nine minutes had been enough for property damage and televised chaos.

He leaned back in his chair slowly.

There were only twelve individuals with access to that clearance override.

One of them was sitting two offices away.

Kandjii closed his laptop carefully.

Internal compromise was not theoretical anymore.

It was proximate.

***

Colombian Border, Dusk

Daniela's fever had begun by late afternoon.

The lean villager who had pulled her inside the hut, Mateo, he had finally offered, pressed a damp cloth against her forehead with practical detachment.

"You are hunted," he said quietly.

"Yes."

"By who?" He ask.

She closed her eyes. "That's the problem."

The helicopter sounds had passed twice. Once low. Once higher.

Mateo's old radio crackled intermittently on a wooden crate near the door. Spanish news anchors debated oil markets. One mentioned Namibia. Another referenced a leaked corporate advisory network.

Her name was not spoken.

Yet.

She forced herself upright and powered on one of the smaller encrypted drives.

A message had arrived through a dormant relay she had embedded months earlier.

It contained only three words. *Window reopened. Conditional.*

Below it, a location coordinate for the Eastern Mediterranean.

She stared at it. Someone had intervened in the jungle.

Not to save her, but to preserve her as asset. The word no longer angered her. It clarified.

***

London, Orion House

Sir Malcolm did not raise his voice. He never did. The boardroom air felt heavier than usual.

"Windhoek escalation was insufficient," he said calmly.

Adrian Cole shifted in his seat. "The unrest achieved narrative traction."

"It achieved sympathy," Sir Malcolm corrected. "For the President."

A thin silence followed.

Reports lay open across the table, Chinese delegation arrival, Russian diplomatic overtures, and American non-commitment.

Multipolar interference was complicating what should have been a straightforward destabilization.

Sir Malcolm's gaze settled on a different briefing.

Daniela Ruiz, Status: Presumed Neutralized.

He tapped the page lightly. "This assumption," he said, "appears premature."

Cole's jaw tightened. "We are reassessing."

Sir Malcolm leaned back. "Reassessment suggests loss of control." He snapped back in an icy voice,

The younger director from compliance spoke cautiously. "Chinese security alignment could harden Namibian response capacity. If Moscow positions as stabilizer, our leverage narrows."

Sir Malcolm's eyes drifted briefly to the portrait behind him: His great-grandfather besides a drilling rig in a continent that had not been consulted.

"Then we widen the aperture," he said quietly.

"How? Energy markets respond to uncertainty."

Cole understood immediately.

Pipeline sabotage rumours.

Currency pressure.

Credit downgrades triggered by "risk analysis."

Destabilization did not require bullets. Only doubt.

"Proceed carefully," Sir Malcolm added. "We cannot be seen." The room absorbed that instruction without discomfort. Visibility was for governments. Orion preferred influence in the shadows.

***

Langley

Marsh did not like convergences, as separate crises should be managed individually. Convergence meant compression. He studied the intelligence mosaic spread across his secure terminal:

- Chinese security advisory presence expanding in Namibia.

- Russian diplomatic outreach intensifying but carefully non-binding.

- Orion-linked financial instruments shorting Namibian sovereign bonds.

- Daniela Ruiz alive. Probable third-party protective interest.

- Internal Namibian security breach flagged.

It was too synchronized.

He stood and walked to the dark window overlooking the parking structure below.

Washington would interpret this through familiar lenses, China expanding footprint, and Russia opportunistic, African instability cyclical.

But that reading was incomplete.

The destabilizing trigger had originated in corporate space. Private actors had accelerated events beyond state choreography. And now states were responding defensively.

His secure line blinked once. An Istanbul contact. Volkov.

The message was brief: *Instability benefits those who require distraction. Consider that.*

Marsh exhaled slowly. *Iran*. The word rose uninvited.

Daniela's intercepted jungle transmission, *Persian window compromised*.

Another axis. Sanctions architecture.Energy corridors in the Eastern Mediterranean.

If Daniela's leak intersected not only Orion but sanction evasion pathways, then her survival served more than journalism.

It served leverage, and leverage was currency in multiple capitals.

Marsh felt the weight of something approaching systemic shift.

Namibia was no longer just a peripheral test case. It was becoming a power node, and power nodes attracted pressure until something snapped.

***

Windhoek, Late Night

Deputy Director Kandjii stood outside the office of the colleague whose clearance code had delayed riot response.

He did not knock immediately. Inside, lights were still on. Documents were being shredded.

Kandjii's jaw tightened. Internal fracture had crossed into betrayal.

At the Walvis Bay port, Chinese engineers recalibrated security mapping.

In London, Orion prepared financial destabilization instruments.

In Moscow, analysts modelled gratitude curves.

In Washington, restraint thinned.

In a Colombian hut, Daniela calculated her next movement not as reporter, but as bargaining chip.

The lines were no longer parallel. They were converging, and convergence, in geopolitics, rarely diffused quietly.

Chapter 18: Night Watch

Tel Aviv

The room had no windows. Just the low hum of cooling fans, the faint ozone smell of overworked servers, and the cold blue glow of thirty-two screens arranged in three curved rows.

Avi Cohen sat alone in the centre chair, the only light coming from the displays and the small red standby lamp on his console. His face was half-shadow, half-pale skin stretched tight over sharp cheekbones. Thirty-four years old, already grey at the temples. He hadn't seen daylight in six days.

He didn't mind, as the world outside could burn. Inside this room, he controlled the burn.

Screen 14: showed live feed from a drone loitering at 22,000 feet over the Zagros Mountains. Thermal imaging painted the Natanz enrichment facility in ghostly whites and blues. Centrifuge halls buried under concrete and rock. The Iranians had learned after Stuxnet; they'd hardened everything. But not enough.

Screen 7: intercepted IRGC comms, Farsi-to-Hebrew translation scrolling in real time. Panic in the chatter. "The American leak has exposed the pivot. Persian window closing. Accelerate enrichment to 65%. Protect Fordow at all costs."

Avi's lips curled. They were late.

He tapped a sequence. Screen 22 flared: a dark-web mirror of the Caracas-Rostov leak, now mirrored again on Iranian hacker forums.

The Mossad-CIA joint brief on "Pivot to Persia" was already circulating in Persian channels, redacted just enough to look authentic, not enough to hide the truth. Cyber strikes on Natanz, funding for dissident cells, timeline tied to Venezuelan oil security.

Iran's supreme leader had gone on state TV three hours ago, voice trembling with rage, calling it "proof of Zionist-American conspiracy to destroy the Islamic Republic."

Good, Avi thought. Let them rage. Rage makes mistakes.

He opened the asset tracker. Green dots pulsed across Iran: human sources embedded in IRGC, technicians in Natanz, couriers in Qom. One dot blinked red, asset 17-K, a mid-level centrifuge engineer. Arrested at 04:12 local time. Interrogation would start soon. He would talk. They always did.

Avi leaned back, cracked his knuckles. The screens reflected in his eyes like dying stars.

He pulled up the Geneva feed, collapsed talks, U.S. envoy storming out, Russian delegation smirking. Harvey's tweet storm was trending worldwide. Avi didn't care about tweets.

He cared about centrifuges. About the moment Iran crossed the 90% threshold. About the window that was closing faster than Tehran realized.

His console chimed. Priority flash from Langley:

"Leak containment critical. Namibian diplomat Voss now active conduit. Threat level elevated. Authorize escalation on secondary targets if necessary."

Avi's finger hovered over the keyboard. Secondary targets. Windhoek. Elena Voss. The woman who had helped amplify the drop.

He opened her file. Old Texas address, parents still living there. A single photo, crosshairs already drawn.

He didn't hesitate. A command string, six keystrokes. Package dispatched: encrypted instructions to a Mossad-CIA cut-out team in southern Africa. *"Neutralize conduit if she continues. Iran timeline cannot slip."*

The message vanished into the dark net. Avi leaned forward again, eyes locked on the Natanz thermal feed. A new heat signature appeared, small convoy leaving the facility. Engineers? Guards? He tagged it for follow-up drone strike if the order came.

The room stayed silent except for the fans.

Somewhere in the world, empires were screaming.

Here, in the blue light, Avi Cohen waited. Patient. Cold. Ready.

The next fracture was coming, and he would be the one to widen, close or escalate it.

***

Windhoek, Internal Security Directorate

The building had been constructed in the late 1980s, during a period when builders still believed in concrete as deterrence.

The corridors were narrow, fluorescent-lit, and painted in a shade of institutional beige that absorbed warmth rather than reflected it.

At this hour, most offices stood dark, but light bled beneath one door at the far end of the second-floor hallway. Deputy Director Elias Kandjii stood motionless outside it, listening.

Inside, the low mechanical rhythm of a shredder pulsed steadily, paper feeding into teeth, reduced to strips, then silence, then feeding again.

Not routine. Not at this hour.

He felt no surge of anger, only a tightening clarity.

Betrayal inside an intelligence service was rarely ideological; it was usually rationalized through fear, money, or proximity to foreign assurance.

He knocked once. The shredding stopped immediately.

A pause long enough to calculate.

Then: “Yes?”

The voice belonged to Director Samson Uatja, twenty-two years in service, decorated during regional anti-trafficking operations, widely regarded as cautious but loyal.

Kandjii opened the door without waiting for invitation.

The office smelled faintly of burned dust and overheated electronics. On the desk, a neat stack of files had already been reduced to curling ribbons in the shred bin. Uatja stood behind his chair, expression composed, though a sheen of perspiration betrayed strain.

"Working late," Kandjii said evenly. "Reviewing protest analytics, mix with destruction?"

Uatja's eyes shifted briefly toward the shredder before returning to Kandjii. "Sensitive sources."

Kandjii stepped further into the room and closed the door behind him, allowing the latch to click into place with deliberate finality. "The clearance delay," Kandjii said quietly. "Nine minutes."

Uatja's face did not immediately betray recognition, but something in his posture altered, a micro-tightening at the shoulders.

"You authorized it."

"No."

"The override code is traceable."

A longer pause this time.

Kandjii moved closer to the desk, noticing a second device half-hidden beneath a folder, not Government Issue.

Imported hardware. Encryption casing atypical for Namibian procurement.

"Who?" Kandjii asked.

Uatja's voice lowered. "You think this is about protests? It is about access. It is about survival."

The words hung heavy.

Uatja inhaled slowly, as though crossing an internal threshold. “You have seen the pressure,” he said. “China embedding security assets. Russia positioning as guarantor. Western capital destabilizing our currency. Do you believe we remain sovereign in this equation?”

“We remain sovereign if we are not purchased,” Kandjii replied.

A flicker of something like regret passed across Uatja’s face. “It is already too late for purity,” he said.

Kandjii’s gaze shifted briefly to the shredder bin. Among the thin strips of paper, he caught a fragment of text still partially legible.

- *Advisory Retainer, OAS Baltic, Orion Advisory Services.*

“You sold timing,” Kandjii said.

“I bought leverage.”

“For whom?”

Uatja did not answer. The sound that followed was not from the hallway. It came from inside the office.

A suppressed discharge, soft, almost clinical.

Kandjii felt impact before he heard the second shot. His shoulder exploded in heat as the round tore through muscle and buried itself in plaster behind him. He staggered but did not fall, instinct forcing him sideways behind a filing cabinet as a third shot shattered the glass pane above the desk.

Uatja was not the shooter. The window behind him had fractured inward.

A dark silhouette withdrew from the exterior fire escape.

Kandjii drew his sidearm and returned fire toward the shattered window, but the shooter was already gone.

Uatja stood frozen, shock dissolving his composure.

"You brought them here," Kandjii said through clenched teeth.

Uatja's reply was almost a whisper. "I did not think it would be tonight."

Blood soaked through Kandjii's jacket, warm and immediate. As sirens began rising faintly in the distance, internal alert systems triggered by gunfire sensors embedded in the building structure.

Uatja sank slowly into his chair. "They said it would only be pressure," he murmured. "Financial. Political. Demonstrative."

"They always do," Kandjii replied.

Security personnel thundered down the hallway seconds later.

By then, the breach was no longer hypothetical.

It was operational.

***

Colombian Border – Night

The fever had sharpened Daniela's hearing. Every insect vibration sounded amplified. Every helicopter rotor echoed longer than physics should allow.

Mateo had fallen asleep near the doorway, machete within reach. The hut's interior remained lit only by a single oil lamp that flickered against wooden walls.

Daniela opened the encrypted drive again. The earlier message had expanded. A secure channel activated.

This time, text scrolled slowly, as if composed deliberately rather than auto-generated.

You were targeted by non-state contractors. Secondary acquisition team prevented termination.

She stared at the phrasing. Secondary acquisition?

Not rescue. Acquisition. Another line appeared.

The energy leak intersects sanctioned corridors in the Eastern Mediterranean. Persian exposure unacceptable to certain parties.

Persian. Not code, but Iran. Her leak had not only revealed Orion's destabilization networks. It had exposed secondary routing structures, covert energy transfers circumventing sanctions through layered intermediaries, some passing through Iranian-linked entities.

Which meant:

Tehran had stake.

Not ideological.

Financial.

Strategic.

A final line appeared.

You possess material that destabilizes multiple sanction architectures. Your survival is currently advantageous. Further instructions pending.

The cursor blinked. No signature.

But the routing metadata, faint, partially masked, traced through nodes in Bandar Abbas before disappearing into maritime relay networks.

Daniela leaned back against the wall, absorbing the implication.

She was no longer merely inconvenient to Orion, but has become a danger to a multi-billion company, who needed their shadow sanction pathways protected or weaponized.

She pressed her fingers against her temples. Empires did not collide in public first. They collided through supply chains.

***

Tehran, Intelligence Complex

Brigadier Farhad Rahimi stood before a digital map illuminated in muted tones, Atlantic routes,

Mediterranean chokepoints, West African oil basins, and sanction overlays pulsing in faint amber.

An aide summarized quietly. “The Ruiz leak intersects Orion financing and secondary transfer nodes previously flagged as sensitive.”

Rahimi folded his hands behind his back. “And the Americans?”

“Monitoring but non-interventionist for now.” The aide replied.

“And Russia?”

“Observing.”

Rahimi nodded once.

“Iran does not require Namibia,” he said. “But we require disruption of those who threaten our energy flows. The Israelis will this as excuse for war.”

“Ruiz?” The aide ask.

“Keep her alive and try to preserve.”

He turned away from the map. “In a multipolar contest, survival of the right narrative node is sometimes more important than survival of a small government.”

***

Windhoek, State House

President Nadi-Omunongo received the report just before dawn.

Internal security breach. Deputy Director wounded. Director detained pending investigation. External shooter unidentified.

Her face remained composed as she read, but inside, something hardened. This was no longer agitation, as it has become violent, personal and too close to home

She walked to the window, watching the city lighten gradually beneath low cloud.

China had offered stabilization tools.

Russia had offered partnership.

The West was fracturing through corporate overreach manipulating their own governments. Iran was moving quietly through sanction architecture and preparing for war.

And inside her own intelligence service, loyalty had been monetized.

She placed the report down carefully. “Call Colonel Hamutenya,” she instructed her aide. “And prepare a closed session of the National Security Council.”

Her voice was steady, but beneath it ran a current of decision. Neutrality had protected Namibia for the past three decades, but neutrality under assault required teeth.

Chapter 19: Adriatic Crossing

Windhoek, National Security Council Chamber

The chamber was windowless by design, insulated from both sound and spectacle. A circular table dominated the room, carved from a single slab of Namibian kiaat, polished to a dark sheen that reflected the overhead lights like still water.

President Nadi-Omunongo stood rather than sat.

Around her: military command, internal security, finance ministry, and two silent legal advisers.

“We have been probed,” she said calmly. “Financially. Digitally. Operationally.”

No one contradicted her. “Our response will not be theatrical. It will be structural.”

She turned toward Colonel Hamutenya.

“Effective immediately, all foreign security advisory arrangements are suspended pending review. Chinese technical teams will remain limited to port infrastructure.”

A pause. “Simultaneously, we will initiate sovereign audit proceedings on offshore extraction contracts. Full transparency. Public.”

The finance minister stiffened. “Madam President, markets?”

“Markets already move against us,” she replied evenly. Opacity is no longer protective.”

She let the silence settle. “And we open controlled diplomatic channels to both Moscow, China and Washington. Not alignment with either, but dialogue of a non-aligned state.”

The room absorbed the significance. Namibia would not choose a specific bloc, but leverage.

***

Trieste, Italy

Five Days Later

The Adriatic carried a different light than the Atlantic, softer, silvered, and reflective rather than imposing. Trieste's waterfront curved inward like a quiet question, its Austro-Hungarian facades holding the memory of empires that once believed permanence was guaranteed.

Daniela arrived under a different name.

The extraction from Colombia had been indirect, river to truck, truck to airstrip, airstrip to unmarked cargo plane that refuelled twice before touching down in southern Europe. No official passports were stamped. No public records updated.

Andriy and Olena were already waiting in a rented apartment overlooking the port.

When the door opened, there was no dramatic reunion.

Only recognition.

"You look worse," Olena said gently.

Daniela managed a thin smile. "I feel worse."

Andriy closed the blinds carefully before speaking. "You've intersected more than Orion."

"I know." She quietly replied, as she placed the encrypted drive on the table between them.

"This contains sanction bypass architecture tied to Mediterranean shipping insurance, layered through energy subsidiaries. Iranian-linked nodes are embedded in the same financial corridors Orion used for destabilization."

Olena's eyes narrowed slightly. "So everyone benefits from silence."

"Yes," Daniela replied. "Except Namibia."

Outside, a container ship eased toward dock.

***

London, Orion Advisory Sub-Committee

Adrian Cole listened without interrupting as the Middle Eastern consultant on the secure line summarized the situation.

"Ruiz confirmed alive. Location: Northern Adriatic. Secondary actors providing protective custody."

"State?" Cole asked.

"Ambiguous. Russian-linked facilitation probable."

Cole's jaw tightened.

"Recovery?"

"Feasible, but urban environment increases exposure."

"And failure?"

"Undesirable."

Cole leaned back.

"Proceed," he said quietly.

The consultant did not identify his firm.

He did not need to.

***

Trieste

The attack did not begin with gunfire, but with a power fluctuation. The apartment lights flickered once, then stabilized.

Olena noticed immediately.

"Grid interruption?" Daniela asked.

"No," Andriy said softly. "Targeted."

Seconds later, the street below darkened, a localized blackout affecting only three adjacent buildings.

Professional containment.

Olena moved first, retrieving the compact rifle she kept disassembled beneath the bed. Andriy swept the secondary room.

Daniela felt the familiar tightening in her chest, not panic, but inevitability.

The first breach came through the stairwell.

No shouting.

No wasted movement.

The men wore civilian clothing but carried themselves with disciplined economy. Night-vision optics glowed faintly green against their faces.

Not amateurs.

Not ideological militia.

Contract professionals.

The door lock disintegrated inward under suppressed fire.

Olena returned controlled shots, forcing entry hesitation.

A second team moved simultaneously across the adjacent rooftop, attempting lateral access through the balcony.

Coordinated pincer.

Andriy fired through the balcony glass before they fully cleared the ledge.

One figure dropped silently. The remaining attacker adjusted instantly, switching angle without visible emotion.

Daniela crawled toward the kitchen, clutching the drive.

The stairwell team pushed inward with incremental advance, measured, patient.

"Target priority: Ruiz," one voice whispered in accented English over comms.

A flash-bang rolled across the floor. White light detonated the room.

Daniela's ears collapsed into ringing silence.

She felt herself dragged sideways by Olena, who shielded her against the wall.

Gunfire intensified. And then, it suddenly stopped.

Not gradually.

Abruptly.

Different shots now from a heavier calibre.

Short, decisive bursts from outside the apartment.

The rooftop team fell first and then the stairwell went quiet.

Footsteps approached the doorway.

Daniela's vision cleared just enough to register a silhouette framed by emergency stairwell light.

Volkov stepped into the apartment as though entering a negotiation rather than a battlefield.

Two men behind him secured the hallway with silent efficiency.

"Your adversaries are impatient," Volkov observed calmly.

Olena did not lower her weapon immediately. "And you?" she asked.

"I am pragmatic." Volkov said with a wry smile.

He glanced at Daniela. "You have become expensive."

She let out a strained breath. "You were observing."

"Yes."

"And now?"

"Now intervention is strategically justified."

Outside, distant sirens began to rise, local police responding to blackout reports.

Volkov's men moved quickly, retrieving one of the fallen attackers from the stairwell.

"No insignia," Andriy noted.

"Private consortium," Volkov replied. "Israeli-linked security consultancy operating under commercial cover. Contracted through layered energy intermediaries."

"Orion," Daniela said.

Volkov inclined his head slightly. "They prefer deniability."

"And you?" she pressed.

"I prefer balance." As he stepped closer to the table where the encrypted drive lay.

"What you carry destabilizes more than a corporation," he said quietly. "It touches sanction architecture, insurance corridors, maritime energy routing and the shadow workings of a number of countries."

"And you want it."

"I want it not exclusively controlled." Replied Volkov. That distinction mattered.

Outside, police lights began flashing against the building facade.

Volkov's men had already begun extraction protocol.

"We leave now," Volkov said simply.

"Where?" Andriy asked.

"Somewhere quieter and hopefully safer."

Daniela hesitated only briefly before picking up the drive.

Namibia had chosen leverage, and now she would become part of it.

As they moved down the emergency stairwell, stepping past the bodies of men who had been paid to erase her, Daniela understood something with cold clarity:

The era of quiet destabilization had ended. Now, the great powers were stepping forward openly, not with flags, but with interests. And Volkov had chosen to enter the game.

Chapter 20: After the Storm

Windhoek

The rain had arrived with the violence of a seasonal correction, thunder folding over the city, lightning stitching white veins across the Khomas sky, gutters overflowing in the older suburbs where drainage still remembered colonial engineering more than modern planning.

By morning, Windhoek carried that washed, metallic scent that comes only after night storms, dust subdued, jacaranda blossoms flattened against pavement, red earth darkened into a deeper, almost bruised tone.

President Nadi-Omunongo stood barefoot at the window of her private residence wing inside State House, watching water still slide in thin lines down the glass.

The city looked so peaceful, but it was not, as on her desk behind her lay three documents:

- The internal breach report from the intelligence directorate.
- A diplomatic cable from Beijing marked urgent, non-public.
- A summary brief of the Trieste assassination incident.

The President closed her eyes briefly.

For decades, Namibia had survived by refusing to choose, as non-alignment Article 96 was part of their Constitution, by accepting capital from one direction, technical assistance from another, and political recognition from a third, keeping all of it carefully balanced like a pot carried on the head.

But balance requires stillness, and someone had started pushing.

A knock came at the door.

"Madam President," her chief of staff said softly, "the Chinese delegation has landed."

She turned.

"Level?"

"Vice-Ministerial. And an expanded security advisory team."

Of course.

China did not react emotionally.

It reacted structurally.

Walvis Bay was not just a port; it was a hinge.
Uranium was not just ore; it was leverage.
Roads were not just asphalt; they were corridors of commitment.

Billions lay embedded in Namibian ground under Chinese contracts.

If she fell, if instability came, insurance recalculations alone would cost them more than diplomacy ever could.

"Prepare the secondary reception chamber," she said. "No flags. No photographers."

This would not be theatre.

***

Windhoek State House

Vice Minister Liu Wen entered with the quiet precision of a man who believed in long horizons.

He was younger than she expected. His hair was threaded lightly with grey at the temples. His handshake was firm and respectful.

Behind him stood two men who did not speak. They wore civilian suits but moved with the contained alertness of trained protectors.

"Madam President," Liu said, inclining his head slightly. "We trust you are well."

"Well enough," she replied.

Tea was served. Neither of them drank it immediately.

"We are observing increased volatility surrounding Namibia," Liu began, voice smooth, almost conversational. "Financial pressure. Information leakage. Security incidents."

He did not say *Trieste*. He did not say *Orion*. He did not say *foreign contractors*.

China rarely spoke the name of the blade; it described the wound.

"Our investments here are long-term," he continued. "Instability benefits no one."

"And stability?" she asked quietly.

"Stability benefits those who build patiently for the future."

There it was. Not a threat. An equation.

Liu opened a slim folder and slid it toward her.

Satellite overlays of Walvis Bay.
Cyber intrusion attempts traced and neutralized.
A technical proposal for perimeter security augmentation around State House and key energy infrastructure.
Temporary advisory deployment.

She looked at the pages without touching them.

"You understand," she said carefully, "that Namibia will not exchange one form of dependency for another."

Liu met her gaze without flinching.
"Dependency is inefficient," he replied. "Mutual interest is more sustainable."

Silence settled between them, thick but not hostile.

Outside, somewhere beyond the compound walls, traffic continued. Life continued. Vendors continued selling fruit at the roadside opposite the American Embassy.

History often turns while ordinary life remains unaware.

"If I accept augmentation," she said, "it remains under Namibian command authority."

"Of course."

"And no public acknowledgment."

"Agreed."

They both knew that satellites would notice anyway.

But public acknowledgment shapes narrative, and narrative shapes reaction.

She closed the folder gently. "We will cooperate."

Behind Liu, one of the silent security men shifted his weight almost imperceptibly, the faintest sign that the equation had tilted.

China had stepped closer.

Not to dominate.

To ensure she did not fall, as the Namibian ship needs to stay stable in this geopolitical storm, for too much was invested.

***

The Speech Draft

The speech lay on her desk like a verdict.

She had rewritten it three times.

The first version was cautious.

The second was defiant.

The third was honest.

She removed entire paragraphs that sounded like strategy documents. She inserted lines that felt personal.

When she finally stood at the podium hours later, she carried no papers.

"My fellow Namibians," she began, and the words felt heavier than they ever had before.

The cameras captured her face in clean, unforgiving light.

"For many years, we have believed that neutrality alone would protect us."

A small pause.

"Neutrality is valuable. But neutrality without resilience invites pressure from foreign entities."

In London, screens flickered.
In Beijing, analysts leaned forward.
In Washington, aides exchanged glances.

She did not accuse.
She did not dramatize.
She described.

Unusual currency volatility.
Foreign advisory influence.
Coordinated security anomalies.

Then she said the words that could not be taken back.

"Namibia will implement a Sovereign Energy Alignment Framework, prioritizing transparent contracts, diversified partnerships, and national command over infrastructure security, for we as country has struggled for 24 years to throw off the yoke of colonialism. We will stand proud, for we are the 'Land of the Brave."

The phrase sounded bureaucratic, but said with a deep commitment.

Its consequences were not.

"All offshore extraction agreements will undergo public audit. All external advisory arrangements suspended pending review."

In the chamber, a journalist's pen stopped moving.

She ended not with anger, but with something harder.

"We will engage the world. But we will not be rearranged by it."

The broadcast concluded.

For a moment, there was only the faint hum of studio equipment cooling.

Then the world reacted.

***

Simultaneous, The Release

In newsrooms across Europe and southern Africa, secure files unlocked within seconds of each other.

Daniela had not simply leaked documents.

She had structured them.

Timelines. Payment flows.

Insurance routing. Energy destabilization models targeting the Namibian dollar.

Layered correspondence between Orion-linked advisory entities and private security consultancies.

Not enough to collapse a system, but enough to expose dark intention.

Traders in London stared at their screens as Orion's stock began sliding in pre-market trading.

Insurance underwriters froze secondary approvals linked to southern African offshore projects. Energy futures spiked on uncertainty alone.

In Trieste, Daniela watched the first headlines appear.

She did not feel triumph, just deep exhaustion.

Volkov stood near the window, looking out toward the port.

"It is done," she said quietly.

"No," he replied with a deep frown. "It has begun, as we have opened the Pandora's Box."

***

Windhoek

State House felt different. Not louder, but infinite denser.

Additional vehicles rotated through the perimeter under the pretext of infrastructure coordination. Cyber teams, Namibian and Chinese, worked side by side in a subterranean operations room, their screens reflecting lines of code and satellite feeds.

Protection is never free, but neither is vulnerability.

The President sat alone again before midnight, as the rain had returned, lighter now, almost gentle.

She allowed herself a single thought she had refused to entertain earlier:

"If they come for me next, they will not come quietly."

Her secure device vibrated once.

A message routed through indirect channels:

Trieste stable. Data live. We move carefully.

No signature. She placed the device face down.

Outside, the city slept, but beneath the surface:

Markets recalculated.
Diplomats drafted cables.
Corporations reassessed risk.
Intelligence services updated threat matrices.

Namibia had not chosen a side. It had chosen the high road.

And China, recognizing the cost of collapse, had stepped forward, not as saviour, not as master, but as stakeholder protecting its interests.

The line had been drawn not in anger, but in clarity and clarity is often much more dangerous.

Chapter 21: Dark Streets, Old Ghosts

Windhoek

The Hilton's service corridor smelled faintly of bleach and overheated wiring. Elena Voss pushed through the steel door into the alley behind the hotel, letting it close without sound behind her.

The President's voice still reverberated in her skull. "Double down. Transparency or nothing. Disappear for now. Someone is hunting."

The word had not been metaphorical.

The evening air carried the cool residue of earlier rain. Diesel drifted in from Independence Avenue. Jacaranda blossoms, bruised by storm water, lay crushed against the pavement like spilled ink.

Windhoek after dark could feel deceptively small, a capital that still behaved like a town. Vendors had folded their stalls. Taxis cruised lazily. Office lights blinked out one by one.

Elena did not head toward her flat.

Predictability kills.

She walked south first. Then cut west. Then doubled back north, forcing herself into the rhythm of someone unhurried.

Three blocks later, the feeling came.

Not fear. Recognition.

A vehicle idling too long. Headlights off. Moving only when she moved. And footsteps.

Measured.

Professional.

Matching her pace without closing it.

Her pulse slowed instead of rising. Old training never disappears. It calcifies into instinct.

She turned into a narrow street lined with shuttered storefronts. Corrugated steel reflected the amber streetlamps in dull streaks. A dog barked once in the distance and went silent.

The footsteps followed.

The car rolled around the corner, engine low, predatory.

This was not random crime.

Not street thugs.

This was containment.

She increased her pace, not a run yet. Let them adjust first.

They did.

She broke into a sprint.

The engine revved.

Footsteps accelerated.

She cut right into an alley thick with the smell of rotting fruit and old beer. A toppled crate crunched under her shoe. Somewhere above, a window slammed shut.

The alley spilled into the skeletal remains of the old market square, empty stalls, tarps half-folded, metal frames casting long shadows under broken light.

Doors opened behind her.

Two figures emerged from the vehicle.

Suits.

Earpieces.

No shouting.

One spoke quietly into a cuff mic. The language was not fully clear, clipped, disciplined. Could have been Hebrew. Could have been English. Could have been something else layered in between.

Professional.

Elena slid behind a fruit stall, breath shallow. Her elbow burned from scraping brick. Sweat ran down her spine despite the cold.

Boots on gravel.

Slow sweep. Flashlight beam sliding across concrete.

She held still, as the beam passed and then returned.

She moved. Low. Fast. Toward the railway cut at the edge of the square.

The first suppressed shots snapped through the air like tearing fabric. Sparks flew from metal scaffolding. A bullet ricocheted close enough that she felt the heat of it.

She vaulted the low fence at the railway line and dropped into the dry culvert below. Gravel bit into her palms. Her shoulder screamed.

One of them jumped down after her without hesitation. Trained.

Boots steady behind her.

She ran along the concrete channel, breath loud in her own ears. Above, the car paralleled her path, headlights flashing intermittently through gaps in the railing.

She ducked under a low pedestrian bridge, as the man behind her closed distance, breathing controlled, not frantic.

She stopped abruptly and pressed against the wall.

He rounded the bend.

She lunged.

The collision knocked the air from both of them. They crashed into gravel, rolling. He recovered faster. His forearm slammed against her throat, cutting oxygen.

Her vision began to narrow.

She drove her knee upward, once, twice, into soft tissue. His grip loosened. She twisted, seized a broken chunk of concrete, and brought it down hard against his temple.

The sound was sickening.

He collapsed.

Above, the second man leaned over the railing, flashlight beam stabbing downward. A suppressed round cracked against the culvert wall inches from her head.

She ran deeper into the culvert. At the far end of the channel, a rusted storm drain yawned half-open. She squeezed through, concrete scraping her back, mildew thick in her lungs.

The beam stabbed after her but could not follow.

Voices above. Anger, but controlled.

They would not scream. They would not curse loudly. Professionals rarely do.

She crawled until she reached open air behind a row of modest houses, gardens dark, and curtains drawn.

She did not stop running for another twenty minutes. When she finally leaned against an old pole, hands shaking, she understood with chilling clarity:

This was not intimidation. This was removal.

***

Memory Has Teeth

As she moved through residential backyards, something else rose, not panic, but history.

Texas heat.

Oil rigs against the horizon.

Her father's hands permanently stained black.

Oil runs thicker than family, he used to say.

She had believed him.

Her years of studying Geology at Texas A&M, with fieldwork in Iraq under contract. Maps drawn that were supposed to be neutral.

Geology do not lie, but the people who use the maps do.

Baghdad safe houses. A "consultant" with government credentials who smiled too easily, requests for layered data, subterranean mapping that had nothing to do with extraction.

She had told herself it was still geology.

Then Kyiv.

Maidan.

Gas fields.

Energy independence briefings.

Reports that moved through channels she did not fully see.

She remembered the first time she realized her data had become leverage.

It felt like discovering that your handwriting had been forged onto a confession.

She left. Burned contracts. Came to Namibia. Married a Namibian Game Ranger, and chose red Kalahari sand over black crude.

But shadows do not forget.

The photo of her parents' old house in Texas had arrived two days ago. Crosshairs drawn over the front door.

No signature. Just a message: *We wait for no one.*

Now the hunters had confirmed it.

Somewhere, someone believed she was a loose end, as she retrieved the emergency burner from inside her jacket.

Only one bar of signal. She typed carefully:

Alive. Operational threat confirmed. Foreign professional team. Proceed with AU statement. I go dark.

She did not speculate about who. Speculation is emotion. Intelligence is restraint.

She sent it, and then smashed the phone under her heel. The viper's nest did not belong to one serpent.

And she had once mapped its tunnels.

Chapter 22: Long Way North

Windhoek

Elena did not rest as she moved north toward the industrial rail yard, where sodium lights cast the world in artificial orange and security was more concerned with cargo than faces.

She kept her posture relaxed. Fear attracts attention.

At the edge of the freight yard, a cargo train idled getting ready for the coast, metal wheels ticking softly as heat dissipated.

Slow steady movement is camouflage.

She slipped between cars and climbed into an empty livestock container. The metal smelled of old animals and rust.

She had two water bottles, a torch, and an encrypted phone with a battery at thirty-one percent.

When the train lurched forward, she did not look back at the city.

She had learned that particular discipline in her first year at Orion, when she still believed she was building something. You do not look back at what you are leaving. Looking back is the beginning of hesitation, and hesitation, in her experience, was how people got caught.

***

Walvis Bay

The coast felt colder than it should have. Fog rolled in low and stubborn, swallowing cranes and container stacks whole. Chinese characters were stencilled across several shipping containers near the port's newer terminals, quiet reminders of who had financed which expansion.

She avoided the main gates. Slipped through a rusted gap in the harbour fence where the concrete footing had subsided, pulling the chain-link just wide enough.

Found a trawler loading under minimal supervision.

Cash still speaks in ports.

The Angolan captain did not ask questions. He looked at the money, looked at her once, then pointed to a narrow space below deck between net coils and diesel drums.

At sea, she lay on her back staring at the corrugated metal above her as the vessel cut north along the African coast. The engine noise was constant and oddly settling. She ran through what she knew.

Orion had been exposed. Namibia had declared posture. China had stepped closer. Russia had moved from observation to intervention.

When systems are exposed and embarrassed, they retaliate. Not publicly. Clinically.

She had worked inside that logic for seven years. She understood its grammar.

That understanding was the only advantage she still had.

***

Lisbon

Portugal felt different. Older. Less forgiving.

Her passport was scanned without comment at arrivals. Her face did not match any active watch list, or if it did, the alert had not propagated quickly enough.

She bought coffee at a kiosk and sat where she could watch the exit doors. The habit was automatic now.

From Lisbon she moved by rail. Madrid, then north. Each city, she altered small details. Hair tied differently. Glasses from a pharmacy kiosk. Jacket replaced at a charity shop on a side street in Salamanca.

She bought a burner in Madrid.

In a café bathroom, beneath the hum of fluorescent light, she typed one message:

Alive. Moving north. Data secure. Intersection point?

The reply came within minutes.

Zurich. Window narrow. Watch for surveillance.

No names. No sentiment. Clean.

She destroyed the phone and dropped the pieces in separate bins across three streets.

***

Overnight train, Spain to France

The landscape shifted outside the window from Iberian brown into the early shadow of the Pyrenees.

She allowed herself a few hours of sleep, hat pulled low, bag between her feet.

When she woke, the train had crossed into France. The light was grey and level. A family in the seats across the aisle ate sandwiches and argued quietly about directions.

She watched them without appearing to.

She had once helped corporations read the earth, mapping its hidden reserves, quantifying what lay beneath nations that didn't yet know what they were sitting on. She had written the reports that made Orion's boardroom lean forward. She had believed, for longer than she should have, that expertise was neutral. That analysis was analysis.

It was not.

Every number she had produced had been a tool in someone else's hand.

The train moved north. Somewhere ahead, in a city built on neutrality and hidden vaults, threads were converging. Daniela. Volkov. Andriy. Olena.

The exposure had begun. But the retaliation was only starting.

She rested her head against the glass and let the motion carry her toward what came next.

Chapter 23: All Paths Lead to Zurich

Moscow

Snow had fallen all night, not in soft flakes but in hard, wind-driven sheets that scoured the avenues and collected in the gutters like ash from an unseen fire. Moscow in winter does not invite mercy; it enforces discipline. Every step sounds deliberate. Every shadow feels inhabited.

The old ex-KGB Colonel has quietly called him back to Moscow, as some information could only be shared face-to-face.

Dmitri Volkov moved through the near-empty streets with the contained composure of a man who had already crossed a threshold he could not uncross. His collar was turned up against the cold, but his eyes were uncovered, scanning reflections in shop windows and the muted gleam of headlights passing at distant intersections.

Two hours earlier, in an underground parking garage beneath an administrative building near the Garden Ring, the illusion of internal stability had collapsed.

He had known the choreography the moment the first man stepped out from behind a concrete pillar. The spacing between them had been correct. The silence had been correct. One held a suppressed pistol low and ready; the other carried a knife not as intimidation, but as certainty. The knife meant that if the pistol failed, proximity would finish the work.

There had been no words exchanged.

Volkov had drawn and fired almost simultaneously, the first shot entering beneath the sternum before the assassin had fully adjusted his stance. The second man had closed the distance faster than expected, the blade grazing Volkov's forearm before a close-range shot ended the motion. The sound inside the concrete chamber had been muted but final.

When security patrols eventually descended, they would find two bodies and a question that would never be publicly answered.

He understood what the attempt meant.

No formal order had been issued through official channels. There would be no paperwork. This had come from a faction, men who believed the world had already crossed into open confrontation and that hesitation was treason. To them, Volkov's preference for controlled destabilization, for calibrated pressure rather than spectacle, had become softness.

Softness, in certain circles, is fatal.

He turned into a narrow courtyard off Arbat Street, where older brick buildings enclosed a space insulated from the wind. A metal door without markings stood beneath a single exterior lamp. He knocked three times, waited, and then heard two measured knocks from inside.

The door opened.

The safe house smelled of strong black tea and aging paper. Heavy curtains blocked the windows. At a plain wooden table sat the man Volkov still referred to as the Colonel, though his official KGB rank had long since dissolved into something much bigger and grander as leader of a nation. Age had thinned his hair, but did not diminished the steadiness in his gaze.

"They moved sooner than expected," the Colonel said without preamble, his pale blue eyes cold.

"Yes," Volkov replied.

"You were warned that the hardliners would not tolerate ambiguity indefinitely."

Volkov removed his gloves and placed them on the table. "Ambiguity prevents escalation into a nuclear war."

"It also frustrates those who prefer clarity from war or direct conflict."

The Colonel slid a passport across the table. Swiss. Clean. The photograph was recent but not widely circulated.

"One way," the Colonel said.

Volkov studied the document without touching it. "And after?"

"There may not be an after, as you have already earned your retirement, finish the task and then you decide."

The words were delivered without drama. In Moscow, drama is wasteful.

"They want rupture," Volkov said quietly. "A single act large enough to force alignment."

"They want the world divided cleanly," the Colonel replied. "They believe division strengthens resolve."

"And Zurich?"

"Neutral ground," the Colonel said. "Where old channels still exist. Where decisions can be made without cameras."

Volkov finally picked up the passport. The paper felt heavier than its weight suggested.

"They will follow me."

"They already have, and I will take care of it," the Colonel answered. "One team remains here to erase traces. Another waits in Switzerland. Their mandate is simple: protect you, silence the internal leak, and ensure blame travels west."

Volkov allowed himself the faintest, almost weary smile. "Then I must arrive before their narrative does. While they play checkers, there is game of chess to be played."

He left through the back entrance. A black Volga waited in the courtyard, engine idling, driver silent. They drove through streets still coated in fresh snow toward a private airstrip on the outskirts of the city. The small jet stood ready, lights off, crew minimal.

As the aircraft lifted through low cloud, Moscow receded into a field of dim lights swallowed by white. Volkov rested his hand briefly over the pistol beneath his coat.

He had spent his career preventing total war through calibrated manipulation. Now he was being pushed toward a decision that might either preserve balance or shatter it completely.

Zurich would not be a refuge for any of the rag-tag group, but at least he could prepare for the reckoning coming from exposing the sinister facts.

In Zurich, Andriy powered on their final burner phone. The battery icon flickered, uncertain.

A message appeared. *En route Zurich. Data safe. Meet if possible.*

Elena. Alive. Olena closed her eyes briefly. “She made it out.”

“For now,” Andriy replied.

She studied his face. “And Volkov?”

Andriy hesitated. “His intervention came too precisely. That troubles me.”

“It troubles me as well,” she said softly.

***

Zurich

Snow rested lightly along the Limmat River, its surface broken only by the steady movement of trams and the measured pace of pedestrians wrapped in dark winter coats. Zurich wore neutrality like a tailored suit, precise, understated, and unassuming.

Volkov arrived through a private terminal, passing through security channels designed for discretion rather than spectacle. He stepped into the cold

Swiss air aware, almost immediately, of the subtle shift in rhythm around him.

A sedan that lingered half a second too long before merging. A pedestrian whose gaze tracked him not directly, but through reflection.

The faint delay in a traffic light that should have changed sooner.

Surveillance in Europe does not announce itself. It arranges itself.

He entered his hired car without haste. Within minutes, two vehicles settled into loose formation behind him, far enough not to alarm, close enough not to lose.

Elsewhere in the city, Elena emerged from Zürich Hauptbahnhof and felt something similar, a thinning of coincidence. A man with a newspaper who never turned a page. A woman who adjusted her scarf repeatedly while remaining within peripheral distance.

She altered her route twice, then a third time, testing response.

The pattern held.

They were not improvising.

They were waiting.

Zurich had become a magnet for intersecting interests.

In a rented apartment near Seefeld, two men unpacked equipment with practiced efficiency. One monitored encrypted communications on a secure laptop; the other assembled a compact rifle calibrated for urban engagement.

Their instructions were simple and unadorned: remove Volkov and his associates, disrupt convergence, ensure attribution misdirected westward.

Across the river, in a parking structure beneath a financial services building, another team reviewed transport schedules and station footage. Their mandate differed only slightly: contain the leak's human vectors, prevent coordinated exposure, avoid spectacle.

Multiple hunts had begun, each for a different prey.

As afternoon light faded toward evening, the geometry tightened.

Volkov exited his vehicle near the Bahnhofstrasse, aware that both trailing cars had adjusted position. He walked deliberately, neither accelerating nor slowing, allowing those behind him to grow comfortable in their rhythm.

Several blocks away, Elena crossed the Münsterbrücke, boots crunching lightly against compacted snow. She sensed compression in the air, an invisible narrowing.

The first shot shattered a tram window thirty meters ahead of her. Glass exploded outward in a spray of glittering fragments. Passengers screamed. The tram screeched to a halt.

The shot was not meant to kill.

It was meant to rearrange.

A second report echoed off stone façades, closer this time. Pedestrians scattered. Sirens began to wail somewhere beyond the river.

Volkov heard the sound and stopped walking.

He did not hesitate.

He turned toward it.

Zurich, city of secret vaults and old agreements, had just been pulled into the open.

All paths had led here. And none of them promised safety.

Chapter 24: Storm Warnings

Global News Wires

The first tremor did not come from missiles.

It came from screens. Across continents, devices flickered and refreshed, analysts leaned closer to monitors, diplomats paused mid-sentence as notifications stacked on top of one another. The leak, Namibia's alignment, the exposed energy manoeuvring, the quiet implication of deeper strategic games, had begun to metastasize. What had started as a regional assertion of sovereignty was now mutating into something far more combustible.

In Tehran, winter air pressed low over Azadi Square, thick with exhaust and human heat. Tens of thousands gathered beneath towering screens bolted to scaffolding, their breath rising like incense into sodium-lit haze. When the Supreme Leader appeared, the camera angle deliberate and low, the crowd's murmur tightened into a unified pulse.

His voice was not hurried. It did not need to be.

"The Zionist-American axis reveals its fangs," he said, each word measured, amplified, and allowed to settle. "Venezuela's oil was the prelude. Cyber-sabotage at Natanz. Proxies stirring unrest. Economic siege dressed as diplomacy. The Islamic Republic does not mistake coincidence for innocence."

The crowd erupted, not in chaos but in rhythm. Chants rolled outward, synchronized, and familiar. Flags snapped against steel barricades. Somewhere in the square, a young man live streamed the speech to millions more who would consume it in kitchens and dormitories and military barracks.

Within hours, smoke was reported rising from a peripheral facility near Natanz. Iranian state media spoke of "external aggression." Western intelligence briefings described "an industrial malfunction." No one believed either version entirely.

In the Persian Gulf, missile tests were announced as "defensive demonstrations." American carrier groups adjusted position. Satellite feeds captured the subtle choreography of escalation, ships shifting grid patterns, aircraft repositioned, and radar signatures brightening and fading like nervous heartbeats.

In Washington, the Situation Room felt smaller than usual.

The President stood before a wall of live feeds: Tehran crowds, Baghdad protests, Tel Aviv analysis panels, scrolling economic forecasts already predicting market turbulence by Asian opening time.

An aide handed him a preliminary intelligence digest highlighting "Namibian amplification of hostile narratives."

Another briefing warned of "increased Iranian rhetoric leveraging African alignment."

He skimmed, jaw tightening.

On his secure terminal, the cursor blinked in an empty tweet field. He hesitated, aware that every word would ripple across allies and adversaries alike. The first draft was shorter, sharper. He deleted it. The second was louder.

IRAN PUSHING FAKE STORIES. LEAKERS & LIARS WILL PAY. U.S. MILITARY READY. DON'T TEST US. NAMIBIA, STOP SPREADING PROPAGANDA OR FACE CONSEQUENCES. AMERICA FIRST.

He hit send.

The room did not exhale, but the world exploded.

Across the Atlantic, in a discreet office suite overlooking Zurich's Bahnhofstrasse, representatives of an old energy empire watched the tweet scroll across multiple screens at once. The firm, generational, aristocratic in posture, burdened with decades of environmental litigation and human rights accusations quietly settled in back channels, did not respond publicly. It never did.

But inside the boardroom, where polished wood and oil portraits of former chairmen stared down like ancestral judges, the atmosphere tightened.

"Containment is slipping," one director murmured.

Another adjusted his cufflinks, gold reflecting soft recessed lighting. "Namibia was supposed to be manageable."

A third, older, with the weary arrogance of inherited power, spoke without raising his voice. "It still is. Political systems wobble. They do not collapse without encouragement."

No one said the name Orion aloud. They did not need to. The network of contractors, consultants, and deniable intermediaries operated like capillaries beneath the visible skin of statecraft.

In Tel Aviv, a secure line hummed between senior intelligence officials and corporate liaisons. Official denials had already been drafted regarding Natanz. Official silence regarding Africa would hold for now. But pressure points were being catalogued: uranium flows, trade exposure, vulnerabilities in small democracies that relied too heavily on Western banking systems.

In Beijing, the tone was different.

At a high-rise near the Third Ring Road, policy analysts reviewed satellite overlays of Walvis Bay port and uranium extraction sites in Namibia's Erongo region. Billions in Belt and Road investments did not sit comfortably beside instability. Roads carved through Namibian desert plains, rail corridors linking mines to harbour cranes, infrastructure built with Chinese steel and financed by Chinese credit, these were not symbolic gestures. They were arteries and investments.

A senior official tapped a pen against the table. "If Windhoek destabilizes, logistics destabilize."

"And if the President falls?" another asked.

The answer was simple.

"Unacceptable."

***

Windhoek, State House

President Nadi-Omunongo stood alone for several minutes before returning to her desk.

The lights below shimmered against damp streets, Windhoek momentarily tranquil despite the digital storm building beyond its skyline.

On her desk lay the African Union draft statement. It was careful language, too careful. Phrases like "transparency in resource governance" and "responsible international conduct" softened accusations without erasing them. The document avoided naming states outright. It spoke of systems, pressures, and destabilization.

Her phone vibrated.

First: the U.S. embassy requesting "clarification."

Second: a message from Addis confirming emergency AU consultations.

Third: a secure diplomatic line flagged from Tel Aviv.

She let it ring once more before answering.

The Israeli ambassador's tone was courteous, almost warm.

"Madam President, we urge restraint. The region is volatile. Misinterpretations could damage long-standing mutual interests, especially regarding uranium trade channels. Stability benefits everyone."

She listened, gaze fixed on the city lights.

"Stability," she replied evenly, "is not preserved by silence."

When the call ended, her aide entered quietly.

"Madam President, Washington is hinting at trade repercussions. Aid reviews. Visa reviews."

"And Beijing?" she asked.

"Beijing has requested a private conversation tomorrow morning."

She allowed herself the faintest nod. China would not threaten openly. It would frame protection as partnership. But she understood the subtext. Investments of that scale did not tolerate unpredictability.

Another briefing lay open on her desk, a fragment recovered from the leak. It referenced "leveraged compliance frameworks" and "coordinated pressure across peripheral democracies." The language was clinical. The implications were not.

She thought of Namibia's liberation history. Of choices made under heavier shadows than tweets and sanctions. Of how easily small nations were expected to fold when giants cleared their throats.

"Send the AU draft," she said finally. "With our amendments intact."

Her aide hesitated only a fraction. "Yes, Madam President."

As the statement moved across diplomatic wires, the reactions were immediate.

U.S. embassy staff in Windhoek began "temporary relocations." Israeli channels warned of "long-term consequences." Markets twitched. Analysts on global networks began speaking of "an African pivot in the new energy cold war."

In Tehran, state television replayed Namibia's language as validation.

In Washington, advisors debated whether Namibia's stance required containment or dismissal.

In Beijing, contingency plans for "protective advisory deployment" were quietly accelerated.

And in quiet offices in Zurich, London, and Houston, men who had never set foot in Windhoek, calculated whether a small republic could be nudged back into their sphere of alignment, or whether a sharper lesson would be required.

Back at the window, President Nadi-Omunongo watched a lone vehicle move through the wet streets below. Ordinary citizens heading home. Shop lights flicking off one by one. A city unaware of how tightly it was now threaded into global tension.

Empires did not bark without reason.

But neither did they retreat without cost.

She turned off the office light and let the rain-slicked reflection of Windhoek hold her gaze a moment longer, aware that somewhere beyond the horizon, forces were already moving, some to protect her, some to replace her, and some to erase her entirely.

The storm had not yet broken, but the pressure had become an unmistakable ticking bomb created by Namibia's oil discoveries.

Chapter 25: Shadows Converge in Switzerland

Zurich

Snow fell lightly over the Altstadt, softening the edges of medieval facades and muting the sound of passing trams. The Limmat River moved dark and indifferent beneath arched bridges, reflecting yellow streetlamps in fractured ribbons. Church bells marked the quarter hour with quiet certainty. Bakery windows glowed warm against the cold.

Zurich believed in discretion. In neutrality. In the quiet management of other people's storms.

It had not yet realized that the storm had arrived.

Elena Voss entered through the alley behind the bakery just before ten. The air smelled faintly of yeast and sugar beneath the sharper scent of snow. She paused beneath the fire escape, scanning the rooftops and the mouth of the alley, watching for the stillness that sometimes meant surveillance.

Nothing moved.

Inside, the safe house apartment was spare, two cots, a folding table, maps, burner phones, a small propane stove. The radiator ticked softly. Through the narrow window she could see the river sliding past in black silence.

She locked the door, then leaned her forehead briefly against it.

From Africa, and now here.

Her ribs still ached where she had struck the steering wheel days earlier. The bandage beneath her sweater was stiff with dried blood. She caught sight of herself in the bathroom mirror, older than she remembered, eyes harder, something burned away that had once believed exposure alone was enough.

Namibia had stepped into the light, and the world had answered with teeth.

A knock came twenty minutes later, three slow taps, pause, two quick.

Andriy.

Elena opened the door carefully.

Andriy Kovalenko stood in the dim hallway, shoulders hunched against the cold. His coat was too thin for Swiss winter, his arm wrapped in fresh gauze that had already begun to darken through. Behind him, Olena Marchenko watched the stairwell, one hand inside her jacket, eyes scanning angles before she stepped in.

No one embraced.

They entered like soldiers entering a contested room.

Olena checked the window sightlines first. Andriy bolted the door, then leaned briefly against the wall as if recalibrating his balance. Only then did he meet Elena's eyes.

"You brought it this far," he said quietly.

"I didn't envisage the consequences," she replied.

No one argued the point.

They stood in the small room listening to the city breathe below them, the hum of distant traffic, the faint rattle of dishes in the bakery kitchen beneath. A city unaware that its calm geometry now hosted converging fault lines.

Midnight came without ceremony.

The knock that followed was wrong. Too fast. Uneven. Desperate.

Olena's pistol was drawn before the second impact hit the wood. Andriy moved to the side of the door, back against the wall, weapon ready despite the tremor in his wounded arm. Elena peered through the narrow camera feed they had rigged earlier.

Daniela Ruiz was barely upright.

When the door opened, she collapsed inward, snow and blood marking the threshold. Her breath came shallow and ragged, fever bright in her eyes.

"They tracked me," she whispered. "Train… Lyon… I had to jump."

Her coat was torn at the shoulder. A thin line of blood traced down her temple into her collar. Her hands clutched a backpack as if it were the last stable object in the world.

Elena knelt beside her. "You're safe."

Daniela shook her head faintly. "No one's safe."

They eased her onto a cot. Olena moved efficiently, cutting away fabric, assessing wounds. Andriy watched the hallway through the peephole.

When Daniela finally spoke clearly, the room grew smaller.

"My mother and brother," she said. "They're here. In Zurich. Industrial district near the lake. They're holding them as leverage."

The words did not rise in panic. They fell in certainty.

"They knew you'd come," Elena said.

Daniela nodded. "They always knew."

She unzipped the backpack and placed the encrypted drive on the table. Beside it, a folded paper with coordinates, surveillance notes scribbled in cramped handwriting.

"If I give them the data, maybe they let them go. But if I don't…"

"They won't," Andriy interrupted quietly.

Silence settled again.

Outside, a tram bell chimed faintly across the river.

Olena unfolded the map. "We scout first," she said. "We don't rush into bait."

Daniela's jaw tightened. "There isn't time."

"There is always time to avoid dying," Olena replied.

They moved at two in the morning.

Zurich's old town lay nearly empty, fog gathering along the water's edge. The industrial district near the lake felt detached from the postcard centre, low concrete buildings, chain-link fencing, and floodlights casting sterile pools onto frozen asphalt.

The warehouse stood squat and functional, surrounded by wire topped with razor loops that glistened faintly in the light. A delivery truck idled briefly at a distant intersection before turning away.

Andriy circled the perimeter first, steps deliberate, breath controlled despite the pain radiating down his injured arm. Two guards near the main entrance. Camera coverage wider than expected. Too professional for simple contractors.

Inside, Daniela's pulse pounded so loudly she feared it would echo across the gravel.

They climbed to the roof where snow muffled their movement. The metal groaned softly under shifting weight. Olena pried open a ventilation hatch with careful pressure.

The air inside smelled of oil and dust.

From the rafters, they saw them.

Daniela's mother and younger brother sat bound to metal chairs beneath a hanging work lamp. Their faces were pale in the harsh light. Fear had hollowed their expressions into something older than weeks.

Four men stood nearby. Suits beneath tactical jackets. Pistols visible but not ostentatious. English murmured in low tones.

"She'll come," one said.

Daniela closed her eyes briefly.

They descended on ropes improvised from straps and cord.

The first guard fell silently beneath Andriy's blade, body eased down without sound. Olena neutralized the second with swift, controlled force. Elena moved toward Daniela's family, fingers working at knots that had cut deep into wrists.

For a moment, it seemed possible.

Then a crate shifted. Wood scraped concrete.

A third guard turned, eyes widening just before he shouted.

The first gunshot shattered the fragile silence of the warehouse. The report echoed violently off concrete and metal, amplifying into disorienting thunder. Dust and insulation rained down from the ceiling.

Andriy fired one-handed, recoil jolting through his injured arm. Pain flared white. A bullet tore through his sleeve, grazing flesh.

Olena collided with a guard in brutal proximity, knife flashing, and the impact driving air from her lungs as they crashed into stacked crates.

Daniela untied her brother fully and shoved him toward the side exit, heart slamming against her ribs.

More men emerged from a side room, too many. Professional. Coordinated.

This had never been just four guards.

Elena grabbed a metal pipe, swinging with raw force into a knee, then again into a forearm. Bone cracked. Someone screamed.

The warehouse devolved into chaos, muzzle flashes, splintering wood, ringing ears. Daniela's mother stumbled as they ran; Daniela caught her under the arms, dragging, whispering. "Just a little more."

A guard stepped into the exit corridor, pistol raised.

The shot that dropped him did not come from any of them.

It came from above. Precise. Suppressed.

The man's body folded without theatrics.

Another guard pivoted upward, confusion registering only briefly before a second round struck clean.

Silence rippled outward in stunned fragments.

From the rafters, a figure lowered himself with controlled efficiency, boots landing softly despite the carnage. Black balaclava. Coat dusted with snow.

He did not rush.

He assessed.

Then, in Russian, almost conversationally: "Me, Volkov. You're slower than I hoped."

Volkov removed the balaclava only once they reached the van a block away. His face was drawn but steady, eyes sharp with something that was no longer simple calculation.

"You were followed," he said to Daniela as he climbed into the passenger seat. "By more than one faction."

Behind them, distant sirens began to rise, unclear whether Swiss police or something else entirely.

As the van pulled away, Andriy pressed his hand against his bleeding arm. Olena scanned rear-view mirrors. Elena drove without headlights for two streets before switching them back on.

Daniela held her mother tightly. Her brother stared at Volkov, unsure whether to be grateful or afraid.

A black sedan emerged briefly in their mirror two intersections back, then turned off.

Above the city, somewhere beyond visible clouds, a surveillance drone adjusted altitude.

Volkov looked ahead through the windshield at snow-laced roads unwinding into darkness.

"In Moscow," he said quietly, "Some have signed my death order, others send me to help."

No one responded. He continued, voice level. "In Washington, they debate escalation. In Beijing, they calculate exposure. In Tel Aviv, they assess leverage. And in Zurich?"

He paused, glancing at the wounded and the rescued alike. "In Zurich, we have chosen sides."

The van merged onto the highway leading out of the city, behind them, the illusion of neutrality dimmed under flashing lights near the warehouse perimeter, ahead of them, Europe lay cold and uncertain.

The storm that had gathered over continents had narrowed now to human scale, to blood on bandages, to mothers held in shaking arms, to men who had crossed lines from which there was no return.

Inside the van, no one spoke. They did not need to, as all paths now had converged.

Chapter 26: Fault Lines

Outside Zurich

The farmhouse sat beyond the last organized geometry of the city, where Zurich's polished facades gave way to fields silvered by frost and narrow roads edged with skeletal trees. Snow moved in fine diagonal sheets across the darkness, driven by a wind that felt older than borders.

Inside, the house smelled faintly of wood smoke and disinfectant.

They had relocated before dawn, abandoning the Altstadt apartment before Swiss police could triangulate too many irregularities. Volkov had arranged the fall-back through a chain of contacts who asked no questions as long as payments were precise.

Daniela's mother slept on a sofa beneath layered blankets, exhaustion heavier than fear for the moment. Her brother lay curled on a mattress near the fireplace, one hand still gripping Daniela's sleeve even in sleep.

Andriy sat at the kitchen table, jaw tight as Olena stitched the torn flesh along his arm with a steadiness that bordered on ritual. He did not flinch. Pain, for him, had long ago become a familiar currency.

Elena worked at the far end of the room, laptop open, cables snaking toward the encrypted drive now fully copied to a secondary device. The farmhouse's router had been stripped of identifying firmware. Every precaution layered atop another.

Volkov stood outside beneath the eaves, collar turned up, and cigarette burning slow between his fingers. Snow gathered along the shoulders of his coat. He watched the treeline the way some men watched horizons in war zones, not expecting movement, but never assuming safety.

Inside, Elena exhaled sharply. "This isn't just operational data," she said.

Olena glanced up without pausing her stitching. "What is it?"

Elena turned the screen toward them.

It was not spreadsheets of oil shipments. Not merely communications about Venezuela or Iran or Namibian leverage. It was modelling.

Scenario trees. Flow diagrams. Political pressure matrices. Corporate data interlinked with governmental shadow operations.

She scrolled slowly.

There were modules labelled by region, West Africa, Eastern Europe, Gulf corridor, Southern Africa. Each contained predictive instability mapping: currency stress thresholds, protest ignition probabilities, leadership vulnerability indices.

Under Southern Africa, Namibia appeared in clean typography.

"Manageable transition framework," Elena read quietly.

Beneath it: subheadings.

Media agitation calibration.Opposition funding vectors.Leadership replacement viability (low-cost scenario).Infrastructure leverage via mineral dependency.

Daniela moved closer, face pale in the blue light.

"This isn't crisis response," she whispered.

"No," Volkov said from the doorway. He had come inside without anyone noticing. "It is crisis design."

The room fell still.

He stepped forward, studying the screen without surprise.

"This is pure destabilization architecture to achieve a specific objective," he continued. "Oil is surface layer. Energy dependency provides narrative cover."

Andriy leaned back slowly in his chair. "You mean they don't just influence governments."

"They pre-build the fall of governments, for maximum exploitation and profit." Volkov replied.

Elena scrolled further.

There were references to private contractor networks. Consulting firms with benign names. Intelligence-sharing agreements routed through unofficial channels. Financial institutions positioned as pressure points.

Some entries were redacted.

Others bore familiar signatures, Orion subsidiaries, liaison offices, inter-agency task identifiers.

Daniela visibly trembled. "If this goes public…"

"It destabilizes markets," Olena said flatly.

"It destabilizes shadow alliances, corporate over-reach in manipulating governments." Volkov corrected.

He moved to the window, staring out at snow sweeping across the fields.

"In Moscow," he said quietly, "hardliners believe destabilization should be visible, tanks, missiles, and borders redrawn. In Washington, destabilization prefers contracts, sanctions and narratives. In Beijing, destabilization is patient, infrastructure first, leverage later."

He turned back to them. "This file shows coordination patterns. Formal agreements between shadow entities, corporate and Western governments. But alignment of outcomes."

Elena understood. If released, it would not simply expose scandal.

It would expose method, facts and manipulation by corporation's destroying countries and populations, all in the name of making a profit.

***

London

In a boardroom overlooking the Thames, the senior directors of Orion Energy convened under muted lighting that did nothing to soften the tension.

The chairman, a man whose family portrait hung in oil behind him, read the preliminary intelligence brief twice before speaking.

"Zurich incident confirmed," he said I an icy voice. "Target assets extracted. Russian defector involvement probable."

A younger executive shifted nervously in his seat. "If the archive reaches full public distribution…"

"It will not," the chairman said coldly.

"Sir, we cannot guarantee that."

The older man's expression tightened, not in fear, but in irritation. "Governments do not allow systemic frameworks to become common knowledge. They manage fallout, and we manage them."

"And if governments are part of the framework?"

Silence.

The chairman removed his glasses slowly. "Then governments will manage it more urgently, with a little push from us."

***

Washington

In a secure room below ground, officials watched satellite overlays of Zurich's industrial district while scrolling through summaries of the Namibian AU statement and Iranian escalation rhetoric.

One adviser spoke bluntly. “If the full architecture leaks, it won’t just hit corporate actors. It implicates multi-administration strategic doctrine.”

A 4-star General leaned forward. “Then we suppress with aggression.”

A State Department strategist shook her head. “Suppression confirms guilt, and we do not have any military assets on the ground in Namibia.”

“And exposure invites fragmentation and even war.” the General replied with an eager look in his eyes.

Another voice, quieter: “China is already expanding protective advisory presence in Windhoek. They frame it as infrastructure security.”

The room tightened perceptibly. “Protective,” the General muttered. “That’s one word for it.”

***

Beijing

In a high-ceilinged conference chamber, the briefing was concise.

Namibia’s president had not wavered as Western pressure was escalating.

Archive exposure risk increasing.

Recommendation: deploy limited security advisory unit under Belt and Road infrastructure protection clause.

Approved.

The directive was signed without ceremony.

***

Switzerland

Back in the Swiss farmhouse, dawn light seeped faintly through frost-layered windows.

Elena closed the laptop halfway, as if reducing the brightness might soften what it contained.

“This is bigger than Namibia,” she said.

"It always was," Volkov replied.

Daniela looked at him curiously. "You knew."

"I suspected," he said. "But suspicion is not proof."

"And now?"

"Now we have proof of structure."

Andriy flexed his stitched arm carefully. "So what do we do?"

Volkov met each of their eyes in turn, and in a very soft voice said.

"There are two paths. One has already started, and could led to our destruction. The other will depend on how strong we are."

He placed a small encrypted drive on the table, separate from Daniela's.

"I still have internal correspondence from Moscow. Hardliner directives. Cross-reference discussions about Western instability operations. If combined with this archive, it forms a complete geometry."

Elena's pulse quickened. "That would be everything."

"Yes."

"And the cost?"

Volkov did not hesitate. "Markets crash. Shadow alliances fracture. Some governments fall. Some wars accelerate."

Silence thickened in the farmhouse.

Daniela's mother stirred softly in sleep.

Snow brushed against the windows like fingertips.

Olena spoke first. "If we release partial exposure, we retain leverage."

"And if we release full exposure," Elena said quietly, "we end our leverage."

Volkov nodded. “Reform,” he said. “Or rupture.”

Outside, the wind shifted direction.

Inside, the architecture of empires lay mapped across a wooden kitchen table in rural Switzerland, not as ideology, not as conspiracy theory, but as interlocking systems of pressure and design to manipulate resources.

For the first time since the leak began, they understood that pressing send would not simply reveal wrongdoing.

It would alter balance.

Elena closed the laptop fully. “We decide together,” she said.

Volkov looked toward the pale horizon beyond the frosted glass. And a though cross his mind.

“Somewhere in Moscow, some players wanted him dead, but if the Colonel’s plan work. Maybe, just maybe, the world could step back from a nuclear war.”

In London, executives calculated containment.

In Washington, strategy warred with fear.

In Windhoek, a president stood beneath gathering foreign shadows.

And here, in the quiet of a Swiss morning, history waited for a decision made by five exhausted people around a kitchen table.

The fault lines were no longer theoretical.

They were active and devastating.

Chapter 27: Protective Shadows

Windhoek

Independence Avenue is bustling with activity, the morning light unusually clear. Windhoek shimmered faintly beneath a sky scrubbed pale and wide. Vendors opened metal shutters with familiar clatter. Minibus taxis idled in uneven rows. Life, as it often did in Namibia, resumed with practical calm.

But beneath that calm, something had shifted.

At Hosea Kutako International Airport, a chartered aircraft bearing no overt military insignia taxied toward a remote apron. The passengers disembarked without ceremony, men and women in dark civilian coats, minimal luggage, eyes scanning not with tourism enthusiasm, but military assessment.

Officially, they were infrastructure security consultants attached to Belt and Road logistics review.

Unofficially, they were not consultants.

By mid-morning, two unmarked SUVs moved through Windhoek's quiet diplomatic quarter, their windows tinted, their drivers disciplined in silence. They did not drive aggressively. They did not signal presence.

They simply existed.

And existence, in geopolitics, is often declaration enough.

At State House, President Nadi-Omunongo stood at her desk, reviewing overnight briefs while sunlight crept across the polished wood surface.

The first memo concerned currency fluctuations, minor, but unusual in pattern.

The second flagged coordinated social media amplification across platforms, most of it originating from accounts registered outside Namibia.

The third, marked discreetly in red, informed her that a small Chinese "technical advisory" unit had arrived to assist with strategic infrastructure security.

She read that line twice.

A knock came.

Mr. Kandjii entered, face measured but alert. “Madam President. The Chinese ambassador requests a private audience at eleven.”

“And Washington?” she asked without looking up.

“The U.S. chargé d’affaires has requested clarification regarding your AU position and foreign security cooperation.”

She allowed herself a small, almost weary smile. “So now they are concerned about cooperation.”

Kandjii did not respond.

Outside the window, acacia trees stirred gently in the wind. Windhoek still looked peaceful. But she had lived long enough to recognize when stillness preceded something less benign.

***

Walvis Bay

Cargo cranes moved methodically against the Atlantic horizon, lifting containers from freighters whose hulls bore the faded paint of global commerce. Chinese engineers in reflective jackets walked the dock perimeter alongside Namibian port officials, clipboards in hand.

Among them, indistinguishable in posture but not in purpose, were two additional figures scanning sightlines, cataloguing access points, noting vehicle patterns.

Belt and Road investment was not merely financial.

It was territorial familiarity. And China did not leave its investments undefended when instability loomed.

Back in Windhoek, murmurs began to gather along Independence Avenue.

The protest had been announced online three days earlier, a loosely framed demonstration calling for "transparency in foreign influence." The phrasing was ambiguous enough to attract broad curiosity.

By noon, several hundred people gathered near Zoo Park. Handmade placards mixed genuine civic concern with strangely uniform messaging. A few slogans appeared in identical fonts, printed professionally rather than painted.

Observers might have missed it.

President Nadi-Omunongo did not.

Inside a nearby café, two men who did not look Namibian watched the crowd through mirrored sunglasses, speaking quietly into encrypted devices. Their accents were European, their posture trained in surveillance rather than agitation.

Money had flowed. Not through obvious channels. Through consultancies. NGOs. "Western democracy accelerators."

Orion did not deploy soldiers. It deployed momentum.

At eleven sharp, the Chinese ambassador sat across from the President in a room that had hosted liberation strategist's decades earlier.

"Madam President," he began, voice calm, "China respects Namibia's sovereignty deeply. Our investments are long-term partnerships. Instability serves no one."

She met his gaze steadily. "Instability is not my objective."

"Of course not. But when external actors seek to manipulate economic pressure or civil unrest, partnership requires preparedness."

He did not say who the external actors were, as he did not need to.

"We are increasing advisory presence at key infrastructure points," he continued. "Discreetly."

"And what would be expected in return?" she asked softly.

He smiled slightly. "Only continued alignment in mutual development."

The language was polite. The implication was structural.

Protection was never free, but neither was vulnerability. .

By late afternoon, the protest had swelled beyond its initial scale.

Chants shifted tone. A bottle shattered against pavement. A small fire was lit near a government building entrance. Police presence increased, measured, restrained, but visible.

Then something changed.

A rumour spread through the crowd, that the President had signed a secret military agreement. That foreign troops were already stationed in the north. The rumour moved faster than verification.

Phones lit up.

Anger sharpened and stones start to flew.

Inside State House, security briefings updated in real time. Drone footage showed clusters of individuals coordinating movement patterns inconsistent with spontaneous protest.

Mr. Kandjii entered again. "Madam President, intelligence suggests funding vectors trace back to European consulting intermediaries."

She did not ask which ones.

"And the Chinese advisory unit?"

"They have offered to assist State House perimeter reinforcement."

She closed her eyes briefly. This was the line.

Allow visible foreign protection, and sovereignty narratives fracture.

Refuse it, and risk escalation beyond control.

Outside, the sound of tear gas canisters popped like distant thunder. Wind carried faint shouts through open corridors.

"Deploy Namibian forces only," she said finally. "Disciplined and measured. No escalation."

"And the advisory unit?"

"Keep them invisible."

Kandjii nodded and exited.

As evening fell, the unrest did not explode, but it did not dissipate either. It hovered, agitated, sustained by something more than organic grievance.

Foreign journalists began to arrive and markets reacted cautiously.

On encrypted channels in London, Orion analysts reviewed footage with dissatisfaction.

"Escalation insufficient," one noted. "Narrative not consolidating."

In Washington, observers debated whether China's advisory presence represented creeping militarization.

In Beijing, a simple line entered a strategic log:

Windhoek stabilization phase initiated.

As night settled over the capital. President Nadi-Omunongo stood alone once more at her office window. Police lights flickered intermittently in the distance. Smoke rose faintly from a burned barricade that had already been cleared.

Her phone buzzed with a secure update:

Currency pressure attempt detected. Temporary banking slow-down in two regional branches. Source under investigation. Economic sabotage layered atop civil agitation.

The architecture she had only glimpsed through fragments was now moving against her.

And yet, State House remained intact.

The protest had not breached the gates.

Infrastructure nodes were quietly reinforced.

Foreign advisors, unseen by the public eye, had positioned themselves at Walvis Bay and along uranium transport routes.

She was shielded. But she was also encircled. Protection from one power. Pressure from another. Manipulation from a third.

Sovereignty had become a balancing act performed above geopolitical tectonic plates.

She turned from the window and returned to her desk, where the AU communiqué and the latest intelligence briefs lay side by side.

In Switzerland, decisions were being weighed that could detonate global systems.

In Windhoek, she was buying time.

Outside, the night air cooled the streets, and the last chants faded into uneasy quiet. The shadows had not withdrawn. They had simply repositioned.

Chapter 28: The Release Decision
Rural Canton of Zurich

The farmhouse was silent except for the faint hum of the refrigerator and the slow, uneven rhythm of wind pressing against the shutters. Snow had stopped sometime after midnight, leaving the fields outside motionless beneath a thin wash of moonlight.

Inside, exhaustion had become a shared language.

Daniela's mother and brother slept in separate rooms now, sedation and shock easing their breathing into something steady. Andriy sat near the fireplace, arm newly bandaged, eyes half-closed but not resting. Olena stood at the window with habitual vigilance, her silhouette framed by frost-etched glass.

On the kitchen table lay two encrypted drives.

Daniela's and Volkov's. Between them, Elena's laptop.

The architecture of empires compressed into metal and code.

Elena had not slept. The final cross-reference process had taken hours, merging Daniela's extraction from Venezuela and Orion's European subsidiaries with Volkov's internal Russian correspondence and intercepted Western alignment notes.

Layer after layer revealed patterns that no single intelligence service would ever acknowledge publicly.

Not collusion in the theatrical sense.

Burt alignment of pressure, shared incentives and clear converging outcomes.

Namibia's destabilization matrix did not sit alone in the files.

There were similar models for smaller states in Southeast Asia. Eastern European pivot points. Gulf and West Africa leverage corridors tied to energy routing and sanction strategy.

The language was sterile, the implications were not.

Volkov leaned back in his chair, watching the screen without touching it.

"This," he said quietly, "is not a conspiracy. It is a habit bred by decades of unchecked power manipulation."

Elena did not look up. "A habit of control."

"Yes."

Daniela folded her arms tightly across her chest. "If we release everything, they will deny it."

"They will try," Volkov replied. "But facts cannot be unseen once mapped."

Olena finally turned from the window. "And markets?"

"They will react. And governments?" Elena said.

Volkov's expression did not shift. "Some will collapse inward. Others will harden."

Andriy spoke without opening his eyes. "War?"

Volkov considered the word.

"Acceleration," he said at last. "Of tensions already present."

Silence filled the room again, not uncertain, but contemplative.

The choice before them was no longer tactical. It was philosophical.

Partial release would expose misconduct without unravelling structure. It would preserve leverage. Allow negotiation. Allow reform to be demanded rather than chaos invited.

Full release would remove leverage entirely.

It would not request reform, but it would force exposure, a reset.

***

Washington

The emergency briefing was no longer framed as "containment." It was framed as "systemic risk."

Intelligence intercepts confirmed that the Zurich cell had consolidated datasets. Russian defector involvement now assessed as "confirmed."

A senior advisor ran a hand across his forehead. "If the Russian internal correspondence pairs with private-sector destabilization modelling, it creates the impression of cross-bloc structural similarity done by the US."

"Which is?"

"Destabilizing."

Another official spoke more bluntly. "It suggests no one is clean."

The room absorbed that quietly.

A cyber-operations liaison cleared his throat. "We can attempt network suppression if we identify distribution nodes early."

"And if the release is decentralized?"

He did not answer.

Because everyone understood the implication.

***

London

In the Orion boardroom, legal counsel had moved from posture to preparation.

"We must assume worst-case disclosure," one attorney said. "Narrative management will be essential."

The chairman sat motionless, hands folded atop the polished table. Generational arrogance had thinned into something more brittle.

"They will not destroy the entire system," he said quietly.

Across from him, a younger director responded, "They might, and we should prepare for plausible deniability."

***

Beijing

The report from Windhoek was stable.

Unrest contained. Currency fluctuation attempts mitigated. Chinese advisory presence unchallenged publicly.

One analyst raised a final concern: "If systemic exposure destabilizes Western leverage structures, ripple effects could impact Asian and African markets broadly."

The response was measured.

"Prepare adaptive positioning."

China did not fear instability. It prepared to benefit from it, as they holds approximately $683 billion to $759 billion in U.S. Treasury securities that could be exploited.

Back in Switzerland, Elena exhaled slowly and closed the final verification window.

"It's complete," she said.

Volkov stood and moved closer. The glow from the screen illuminated the lines carved deeper into his face over the past weeks.

"If we release fully," he said, "we cannot steer the outcome."

Daniela looked at him. "If we don't, we become gatekeepers and targets."

"That is also power," Olena said.

"And power corrupts," Daniela answered, almost reflexively.

Andriy opened his eyes at last. "Not always. But it tempts."

Elena stared at the blinking cursor in the secure upload interface.

She had imagined this moment differently once, heroic, cathartic, and righteous.

Now it felt procedural. Heavy.

"I didn't start this to negotiate better corruption," she said quietly.

Volkov studied her for a long moment.

"In Moscow," he said, "they wanted leverage over escalation. In Washington, they wanted leverage over compliance. In London, leverage over alignment. Everyone believes they can manage fire."

He paused. "Sometimes the only honest act is to remove the illusion of control."

Daniela stepped forward and placed her hand beside Elena's on the table.

"My family paid already," she said. "Namibia is paying and so will many other innocents too."

Olena joined them.

Andriy rose slowly, standing opposite.

Volkov remained still for a heartbeat longer, then placed his encrypted drive beside Daniela's.

"No partial truth, let's do this correctly." he said.

Elena initiated the distribution protocol.

Not to a single outlet.

To multiple across the globe.

Encrypted torrents seeded through decentralized nodes. Independent investigative journalists across continents. Academic transparency networks. Redundant cloud dispersal across jurisdictions hostile to each other.

It would not be easy to suppress.

It would not be easy to trace.

The upload bar began to move.

Five percent, slowly creeping to ten percent.

Outside, the wind quieted.

Twenty-seven percent.

In Washington, an anomaly alert pinged on a network monitoring dashboard.

Forty-three percent.

In London, a cybersecurity consultant frowned at an unusual data burst signature crossing European routing exchanges.

Sixty-one percent.

In Beijing, an analyst paused mid-report as a secure terminal flagged unexpected public archive indexing.

Seventy-eight percent.

Volkov did not blink.

Ninety-two percent.

Elena felt no triumph.

Only release.

One hundred percent upload.

The files propagated outward into a digital ecosystem designed for resilience and replication.

There was no dramatic sound.

No explosion.

Just a final confirmation message.

'Distribution complete.'

For a moment, nothing changed.

Then Daniela's phone vibrated. A journalist in Madrid.

Andriy's burner lit up. An independent outlet in Warsaw.

Olena's encrypted channel flashed with a secure message from a Baltic investigative consortium.

Volkov's device remained silent, but he knew Moscow would already be reacting, as the old Colonel will anticipate the next chess move.

In Windhoek, President Nadi-Omunongo's secure terminal received a preliminary alert flagged by her intelligence unit:

International media surge. Archive leak confirmed. Namibia referenced as destabilization case study.

She read the line twice.

Then she closed the file slowly.

In Washington, voices rose in controlled urgency.

In London, market futures flickered.

In Beijing, analysts recalculated exposure curves.

In Moscow, a senior Duma official used the word "containment" and quickly disappear, as the cold blue eyes of the Colonel drift his direction.

And in a quiet farmhouse outside Zurich, five people stood around a kitchen table, watching screens illuminate across the world as the architecture of influence, once invisible, began to surface in plain sight.

The release did not create the fault lines.

It revealed them. By morning, the world would no longer be arguing about isolated paedophile or woke scandals.

It would be arguing about how weak politicians and corporations manipulate their daily lives, in the name of greed.

And factual truths, once seen, does not easily disappear.

Chapter 29: Persian Pivot
Zurich Safe House

The morning held itself in a brittle quiet: a hush sanded with snow, the brittle light of early winter filtering through tall pines and falling across frost-glazed fields. Inside the safe house, the temperature hovered beneath conversation, as if the air itself suspected the weight of what had been released and was now being awaited.

They gathered again, not from urgency but inevitability.

Elena Voss was at the table before dawn, coffee cold against the hum of the laptop. The archive, once merely data, then revelation, had shown them something they had only sensed before: the deep logic beneath the leaks, the shadow manipulation from which the global conflicts drew their force and destruction.

Above the flickering cursors, she scrolled through files with a deliberation that had become habitual. Lines of strategy, email threads, pressure algorithms and equations of influence, coded in bureaucratic diction but astonishing in structural design.

Andriy watched her from the corner, alert despite sleeplessness. Olena paced, her gaze often drifting to the woodstove as if expecting the flames to speak. Daniela sat silent, clutching a re-warmed mug, her gaze haloed by the screen's blue glow.

Volkov stood by the window, looking out toward snow-kissed fields that hid far more than they revealed.

"It's not what we thought," Elena said finally, her voice low, not accusatory but analytical. "It isn't just scandal or corruption. It's global corruption of systems that could ignite a nuclear war."

She tapped a line of text that brought up a diagram: cascading nodes of energy routes, military deployments, and regional pressures woven into one vast, recursive map.

"What you uncovered was the surface," she continued, eyes not leaving the screen. "This is the underlying doctrine."

Volkov's shoulders did not shift, but his eyes sharpened. "The Persian Pivot," he murmured.

"Yes," she said. "Not a name on a memo. Not a political slogan. A layered strategy."

She pulled up another segment, lines of code and annotations woven through dated email drafts and satellite briefings.

"Look at this," she said. "It connects the Venezuela campaign, the oil leverage there, with the Middle East deployment. And it's not just military. It's about economic, informational, and strategic capturing of resources."

A silence hovered between them, dense with implication.

"In heavens name… why?" Daniela asked.

Elena exhaled slowly.

"It assumes that controlling oil flows enables coercive leverage over global partners and rivals alike. Venezuela was not merely a resource target. It was a pressure point, a pivot node in a larger graph of influence."

Outside, the wind shivered against the eaves.

Volkov turned slightly away from the window.

"And the Middle East?" he asked.

Elena clicked to another chart, one with more recent timestamps: U.S. naval assets gathering in the Arabian Sea and near the Strait of Hormuz, warships circling in layered constellations, larger and more concentrated than at any point since the Venezuelan campaign.

"Look at this build-up," she said. "They compare it to the fleet used before the Venezuela operation, larger, more systematic. The public line is deterrence, a posture to influence negotiations."

She moved to the next line: images of carriers and destroyers, annotated with deployment dates and operational phases.

"But the internal rationale, and we saw this in some leaked strategic briefs, is that this 'armada' is the backbone of a pre-emptive posture against Iran's program, calibrated through energy dependencies and market leverage."

Andriy's gaze hardened. "Are they preparing for war?"

"Not just war," Elena said. "Structural dominance. They believe kinetic force backed by economic control will bring Tehran to heel, and reset the balance under their control."

Volkov remained still, his eyes distant.

"They want energy dominance, military compliance, and geopolitical submission," he said. "Not only of Tehran, but of any state that stands outside their preferred order."

Olena let out a quiet breath. "So it isn't just one conflict," she murmured. "It's a pattern."

"Elites in Washington think they can avoid full-blown war by using calibrated pressure," Elena continued. "Sanctions, blockades, asset seizures, strategic deployments. But that calculus assumes Tehran stays passive."

"That assumption is false," Volkov said quietly, as he pulled up the latest intercepted communications from Iranian state broadcasts, where officials warned that their forces were ready "with their fingers on the trigger," and that they had learned from past conflicts how to retaliate forcefully, a linguistic locus of deterrent threat and survival strategy.

"And behind them," Volkov added, "stand those with whom Tehran has grown close."

He opened a secure image of a trilateral strategic document released separately by allied capitals, one celebrating a multipolar pact between Iran, China, and Russia, an alignment that had jolted global calculations and complicated any unilateral designs.

"This is no longer a Western script alone," he said. "It has become a patient and adaptive counter-architecture."

They gathered closer, the farmhouse walls shaking slightly with a distant gust.

"And Israel?" Daniela asked. Her voice was careful.

"Elimination is their calculus," Volkov replied. "They see Iran as an existential threat. But they lack the capacity alone. They need a partner, a powerful partner."

Elena winced slightly, knowing what that implied. "Which is why Washington's fleet is larger now than anything seen before," she said.

Andriy's eyes narrowed. "So Israel is pulling them in."

"Elites in Tel Aviv believe that pressure must become force," she replied. "Force must become decisive. Force must be definitive."

The room dimmed as clouds scudded across the sun, cutting off winter light.

"Russia and China are assisting Iran not out of sentiment," Volkov said, "but because the balance of energy and strategic weight benefits them too. They have reasons to counter U.S. dominance, reasons rooted in their own contest with American hegemony."

"Shadow fleets transporting Iranian crude," Elena murmured. "Economic lifelines that weaken sanctions."

Andriy stood up abruptly. "So this pivot," he said, his voice taut, "is not a pivot toward peace. It is a pivot toward dominance, through oil, through military force, through regional destabilization."

"Yes," Volkov answered.

"You said the endgame is Iran. Explain. All of it."

Volkov didn't turn, but his voice was soft and clear. "Israel cannot fight Iran alone. Not a full war. Iran has depth, missiles, proxies, and numbers. Tel Aviv knows it. Washington knows it, and the Arab States knows it. So they build the conditions. Quietly."

Andriy leaned forward, arm bandaged but steady. "Conditions?"

"Oil," Volkov said. "Venezuela's oil. The raid wasn't just about Maduro or democracy. It was about control. Flood the market with cheap crude, prices crash. U.S. tightens sanctions on Iran without the world screaming about gas prices. Tehran starves. Proxies weaken. Israel gets breathing room to strike, Natanz, Fordow, missile sites.

But they need American military cover. Carriers in the Gulf. Air support. The CIA and Mossad run the shadow ops, assassinations, cyber, and dissidents. MI6 keeps the distraction alive in Ukraine, tying Russia down so Moscow can't fully back Tehran."

Olena spoke, voice low. "And Russia?"

Volkov finally turned, eyes meeting hers. "Russia supplies Iran with weapons, tech, and oil deals. But the hardliners in the Kremlin want escalation everywhere. They see weakness in the West, opportunity for bigger war. I don't. I see only fire. If the U.S.-Israel pivot succeeds, and Iran falls. Russia loses a key ally.

China loses influence. The fracture becomes a chasm. Nuclear if someone blinks wrong."

Daniela clutched her backpack tighter. "So Venezuela… it's just a distraction for a bigger game?"

Volkov nodded. "The pawn. The oil secures the logistics, frees U.S. assets to move closer to the Gulf. Carriers repositioning now. Venezuela's crude buys time for the Iran strikes. Who is playing who? The CIA thinks they're using Israel to contain Iran. Mossad thinks they're using America to survive. MI6 thinks they're using both to keep Russia bleeding. China and Russia watch, arming Iran quietly, waiting for the West to overreach. Everyone plays everyone. The leak is the only thing that can stop the chain of escalation into a nuclear war."

Elena stared at him. "And you? A Russian FSB officer helping us expose this? Why?"

Volkov's face was stone. "Because I served in Syria. I saw what happens when empires collide, children in rubble, and cities in ash. The hardliners want that again, bigger. I don't.

The leak weakens the West without war. It gives Russia breathing room. It might even force real talks. My interest is simple: survival. Not victory. Pure survival."

"And if Iran decides it must respond," Daniela said, "they could threaten the world's choke-point, the Strait of Hormuz."

Twenty percent of the world's oil exports moved through that narrow waterway, and any blockade there could shatter markets and economies worldwide.

Elena looked from face to face.

"We exposed corporate shadow manipulation," she said. "But this pivot, this structural strategy, is not contained by truth alone."

The wind outside gathered strength.

Inside, they understood that revealing the Persian Pivot had not ended a design.

It had illuminated the battlefield beneath it.

And human reckoning had only just begun.

Chapter 30: First Movements of the Pivot

Global Markets & Hormuz

When the sun rose over the scattered flats of the Persian Gulf that morning, it illuminated a scattering of black hulls and grey decks, aircraft carriers and destroyers forming a slow, intentional circumference around the sea's narrow gateway. It was a choreography of steel and confidence: the U.S. Gerald R. Ford cutting through dawn mist toward waters where every wave carried risk.

But sunlight seemed too casual a thing to hang over a world already shaking beneath strategic tectonics.

***

Ripples in the Financial Sea

In the glass-skinned towers of global finance, traders watched screens not with curiosity but with rising apprehension. Oil benchmarks, the visceral heartbeat of modern economies, had been churning violently in recent days, surging above psychological thresholds and then retreating with maddening speed.

Brent crude jumped again, not because of a new conflict, but because the fear of it was now a structural expectation.

Long-term charts recorded the largest weekly oscillations in months, risk premiums ratcheting higher with every headline hint of tension, every advisory about tankers avoiding Iranian waters, every whispered rumour of how close forces were now arrayed in the Gulf.

Currency desks and bond floors flinched in parallel. Gold crept toward record heights that should have belonged only to the memory of past crises.

Every price line murmured the same lesson: the market had stopped believing in the old certainties. It now traded on fear. It traded on possibility. It traded on the unpredictability of men who believed leverage was the same thing as control.

***

The Chokepoint Looms

Pilots on carriers knew the Gulf was narrow, even claustrophobic. The Strait of Hormuz, barely twenty-one miles at its tightest, handled one-fifth of the world's seaborne oil flows, a narrow bottleneck on which billions of economic lives depended.

Every naval radar sweep, every reconnaissance sortie was not just a military exercise. It was a statement. Doctrines designed in distant offices had moved from theory into geography. Every signal sent across encrypted waves reminded Tehran that power was not abstract; it was material, immediate.

Iran's response was measured outwardly but intense behind closed doors. Speeches from high officials threaded warnings of retaliation not as rhetoric but as strategic calculus. Even the nation's oil sector, already wracked by protests and economic strain, felt every tremor of external pressure.

***

Tehran's Calculus and Unintended Movements

When the news arrived of a downed Iranian drone near a U.S. carrier group, markets jolted again.

Prices spiked, then slid, then held at a loftier baseline than existed weeks earlier. Analysts, once mechanical in their charts, now spoke the language of fear and probability.

Political leaders in Tehran interpreted every U.S. movement not as isolated pressure but as a continuation of a pattern, a pivot that sought to corral Iran into strategic submission by combining naval dominance with energy control. In the crowded corridors of power, they began to speak openly of defence postures that extended beyond bluster and into capability.

Tehran's internal security briefings placed naval positioning, sanctions, and naval advisories not as separate irritants but as a coherent design. It was a worry with teeth.

***

The First Collision Course

The pivot didn't crash into the world all at once. It teased itself out in minor but telling ways.

A tanker diverted at the last minute, its captain disobeying standard routes to avoid controversial waters. Insurance premiums on shipping routes spiked, not because a strike had occurred, but because someone feared one might.

Equity markets paused at each headline, uncertainty spreading from energy stocks into tech and industrial indexes.

And in one global corner, away from high corridors and trading pits, there was a smaller ripple that would matter later.

Namibia's uranium sector, already vulnerable to narrative collapse, saw brief sell-offs tied directly to oil prices and geopolitical risk appetite. Global funds retreated, seeking safe havens, and small capitals around the world felt the pivot in their valuation charts.

It was the first domino. Not a catastrophe. Not yet. But the first unintended manifestation.

***

Diplomacy and Discord

Skirmishes about naval movements and price signals did not preclude negotiation. Envoys met at backchannel tables, in Oman and in Geneva, in half-lit hotel conference rooms where translators hovered outside curtained partitions.

Talks aimed at resuming nuclear dialogues, shared interests, and confidence-building measures unfolded like ghost dances, tentative, uncertain, and shadowed by the larger manoeuvring looming above them.

Yet every diplomatic step seemed accompanied by a logarithm of unease. In the context of the Persian Pivot, diplomacy had become another lever in the larger architecture. Not the antidote to it.

Zurich, Safe House

News updates arrived slow and fragmented, often at odd hours in the still-young European day. Screens blinked with tickers showing oil moves, fleet positions, and financial anomalies.

Daniela watched one screen as it showed Brent hovering near multi-month highs, the lines jagged as instinctive risk premiums rose and fell with the latest bulletins.

"It's not just markets," she murmured. "It's confidence. When energy becomes unpredictable, every system built on it trembles."

Andriy, staring at a map of naval dispositions, did not respond immediately. The mechanics of a pivot, once intellectual abstraction, was now living, breathing in formations of steel and smoke.

Olena stood by the window, coffee cooling in her hand. "They all think the pivot is about dominance. But it's about fear of loss, of control, of narrative, of advantage. Once fear guides strategy, every actor becomes reactive."

Volkov's eyes were distant, as if seeing battles not yet fought. "And fear is now the momentum. Terrible mistakes could be made."

The true first movement was not a strike or blockade. It was a number.

When Brent and WTI moved in unison, it was because traders had calculated, maybe unconsciously, but with brutal financial precision, that the cost of uncertainty was more expensive than the cost of risk itself.

Seventy dollars a barrel was not just a price. It was a threshold. A mirror showing the world what could happen if the chokepoint never reopened, if the naval impasse became protracted rather than transient.

The room fell quiet. Not alarmed, neither relieved. Just aware.

The first real consequence of a doctrine once abstract had arrived in the world's bloodstream.

The pivot was no longer theory. It was happening.

Chapter 31: Fractures in the Doctrine

Zurich

News updates arrived slow and fragmented in the safe house where the Zurich circle reconvened, often at odd hours in the still-young European day. Screens blinked with tickers showing oil moves, fleet positions, and financial anomalies.

Daniela watched one screen as it showed Brent hovering near multi-month highs, the lines jagged as instinctive risk premiums rose and fell with the latest news bulletins.

"It's not just markets," she murmured. "It's confidence. When energy becomes unpredictable, every system built on it trembles."

Andriy, staring at a map of naval dispositions, did not respond immediately. The mechanics of a pivot, once intellectual abstraction, was now living, breathing in formations of steel and smoke.

Olena stood by the window, coffee cooling in her hand. "They all think the pivot is about dominance," she said. "But it's about fear of loss of control, of narrative, of advantage. Once fear guides strategy, every actor becomes reactive."

Volkov's eyes were distant, as if seeing battles not yet fought. "And fear is now the momentum that drives everything. Terrible mistakes could be made."

The room was quiet.

***

The First Domino Falls

The true first movement was not a strike or blockade. It was oil.

When Brent and WTI moved in unison higher, it was because traders had calculated, maybe unconsciously, but with brutal financial precision, that the cost of uncertainty was more expensive than the cost of risk.

$70 a barrel was not just a price. It was a threshold.

A mirror showing the world what could happen if the chokepoint never reopened, if the naval impasse became protracted rather than transient.

In that movement, a number shifting on electronic boards, the Persian Pivot began to show its teeth.

The room in Zurich fell suddenly quiet. Not alarmed, neither relieved. Just aware that the first real consequence of a doctrine once abstract had arrived in the world's bloodstream.

The pivot was no longer theory. It was happening.

***

Washington, D.C

The morning light over the Potomac was thin and reluctant, diffused through winter haze and the glass façades of federal buildings that had seen decades of strategic reinvention. Inside the West Wing, the air felt conditioned not only for temperature but for tension.

The overnight intelligence summary lay printed in thick paper stacks across the Situation Room table.

Oil volatility sustained.
Insurance premiums on Gulf shipping rising.
Iranian naval drills expanded beyond projected parameters.
Russian reconnaissance assets detected in closer proximity to U.S. carrier formations.
Chinese diplomatic messaging emphasizing "multipolar stability."

No single line triggered panic, but together, they formed pressure.

The National Security Advisor read silently for a long moment before speaking.

"We designed the Pivot to stabilize leverage," she said. "Not to accelerate instability."

Across from her, a senior Pentagon official leaned forward, fingers interlocked. "With respect, leverage always risks acceleration."

A CIA deputy shifted in his seat. "The exposure of the clandestine manipulation plan has altered perception. What was once managed through opacity is now interpreted as deliberate dominance."

"And?" the Pentagon official asked.

"And perception shapes response."

Silence settled like dust.

They had not anticipated that the release from Zurich would refract internal doctrine back into public discourse. Editorial boards were now using phrases once confined to classified briefings. Academic forums dissected "energy coercion architecture." Even allied capitals had begun asking careful questions.

The Pivot had been designed as a pressure system, but it was now becoming an uncomfortable narrative.

***

The Pentagon

In a windowless planning room, naval strategists reviewed updated maps of the Gulf. Carrier strike groups remained positioned with disciplined precision, yet the tone had shifted from deterrence to containment.

"We are one miscalculation away," a Rear-admiral said quietly. He did not specify miscalculation by whom or what.

Drone interception reports had grown more frequent. Iranian patrol craft operated just inside internationally recognized limits but with calculated proximity. Russian and Chinese electronic surveillance vessels loitered farther out, technically neutral, strategically observant.

No shots had been exchanged.

But no one in the room believed proximity would remain purely symbolic.

"De-escalation messaging?" someone suggested.

The Admiral shook his head faintly. "If we pull back, it reads as retreat. If we push forward, it reads as provocation."

That was the fracture. Not between hawks and doves, but between confidence and uncertainty.

***

Langley

Inside CIA headquarters, a separate tension brewed, less visible, more corrosive.

Internal assessments acknowledged that elements of private-sector coordination around energy leverage had blurred lines between corporate interest and national doctrine. The Zurich exposure had illuminated interdependencies that were never meant to appear synchronized.

A senior analyst spoke bluntly during an internal review.

"We believed we could modulate markets and regional pressures without systemic blowback."

"And now?" his superior asked.

"Now markets are modulating us."

No one laughed.

The analyst continued, "Iran's alignment with Russia and China has hardened. The Pivot was designed to isolate Tehran. Instead, it has consolidated their partnerships."

Across the table, a policy liaison interjected. "Israel continues to signal urgency. They believe delay strengthens Iran's nuclear posture."

"Which means," the analyst replied, "pressure for kinetic action will increase."

The room absorbed the implication without comment, as actual war was not the objective. But pressure on Iran to conform to US and Israeli demands.

***

Capitol Hill

The fracture did not remain confined to secure rooms.

Congressional hearings began to surface the language of doctrine. Lawmakers, sensing political vulnerability, demanded clarity regarding naval deployments and oil coordination. Questions once dismissed as conspiratorial were now framed as oversight.

A senator from the Midwest held up a printed chart of oil price movements.

"American families are paying for strategic experiments," she said into microphones. "Is the Persian Pivot a doctrine of stability, or a foolish gamble that could end in nuclear war?"

The phrase landed heavily. The administration avoided confirming terminology, but avoidance no longer erased association.

***

Windhoek

Thousands of miles away, Namibian President Nadi-Omunongo read translated summaries of the hearings with measured attention.

She understood something Washington had only begun to confront:

Exposure shifts power not by destroying structures, but by forcing them out of the shadows to reveal themselves.

Chinese advisory presence remained discreet. Infrastructure nodes were stable. Currency pressure had eased temporarily as markets recalibrated.

Yet Namibia's uranium contracts were being renegotiated in the shadows.

Western firms hesitated. Asian buyers leaned forward.

Global uncertainty had increased Namibia's strategic weight, but it had also made her nation a target to geopolitical games.

Mr. Kandjii entered quietly.

"Madam President, Washington has requested a private call."

She considered that for a moment. "Now they want reassurance," she said softly.

"Yes."

"And Beijing?"

"Assurances of continued cooperation."

She stood, walking toward the window overlooking Windhoek's subdued skyline.

"Sovereignty," she murmured, "is most fragile when everyone offers protection."

She turned back. "Schedule both calls. Separately."

***

Zurich Safe House

In Switzerland, snow had begun to melt, thin rivulets carving through the white fields. The safe house felt less like a hiding place and more like a watchtower.

Elena monitored news feeds with analytical detachment. Daniela traced financial movements across continents. Andriy studied naval diagrams with a soldier's instinct for escalation. Olena moved quietly between windows and screens.

Volkov stood apart, listening more than watching.

"They are fracturing," Elena said.

"Yes," Volkov replied. "But fracture does not mean collapse. It means adaptation."

Daniela looked up. "Could they reverse it? Pull back?"

Volkov considered the question carefully.

"They cannot un-see the alignment," he said. "Nor can Tehran. Nor Beijing. Nor Moscow, and definitely not the US."

"And Israel?" Andriy asked.

Volkov's expression tightened slightly.

"Israel fears escalation without clear US support, as it will be the end of Israel as state, but their Prime Minister will push for war, as it is the only thing keeping him out of jail."

The room fell quiet.

Elena leaned back, rubbing her temples.

"We exposed the shadow plan," she said. "But the intelligence architecture resists demolition."

Olena's voice was steady. "Then what changes?"

Volkov looked at each of them in turn.

"Doctrine," he said. "Not immediately. But publicly. They will rewrite propaganda language. Reframe intent. Create distance between state and corporation. Increase false transparency performativity."

"And privately?" Daniela asked.

"Privately, they will recalibrate, and little will change."

The weight of that truth settled heavily.

***

Washington

In the Oval Office, the President sat alone longer than usual after briefings ended.

The Persian Pivot had been conceived first not as aggression but as insurance, a way to secure American leverage of resources in an increasingly multipolar world.

But insurance can become provocation, and provocation can become a nuclear conflict.

Outside, the city hummed with ordinary life, blissfully unaware of how narrowly naval vessels maneuverer near choke points and open conflict, how markets priced in dread, how alliances strained beneath doctrinal ambition.

Inside the halls of power, decisions hovered between restraint and resolve.

***

Zurich

Elena closed her laptop finally. "Nothing has exploded," she said.

"No," Volkov replied. "But pressure has redistributed."

Andriy stared at a map where colour lines marked naval proximity. "So this is what fracture looks like," he said quietly.

"Yes," Volkov answered. "Not collapse, a correction."

Outside, snowmelt dripped rhythmically from the roof.

The Persian Pivot had begun to move in the world not as leverage, but as a build-up to aggression.

Inside Washington, elites questioned its trajectory.

Inside Windhoek, sovereignty tightened its spine.

Inside Zurich, five people watched doctrine adapt in real time.

The first domino had not toppled into war. It had toppled into doubt and recalibration.

Chapter 32: The Tightening Net

The first sign that Zurich was no longer neutral came without sirens. It came as a digital silence.

Daniela noticed it when her banking app refused to load. A bland message appeared where her balance should have been, *temporary review in compliance with international financial protocols.*

It was phrased politely, almost apologetically, as if global power were a customer-service inconvenience.

Within an hour, Andriy received a similar notification. By afternoon, Olena's residency renewal appointment was postponed "pending clarification."

Volkov said nothing when his secure satellite channel failed to authenticate. He simply looked out of the window of the rented apartment above the Limmat River and measured the flow rate.

Pressure had become coordinated.

Swiss neutrality, that quiet cathedral of discretion, had begun to tilt under shadow weight.

Diplomatic cables moved through Bern with the density of a winter fog. Moscow had filed a formal extradition request for Viktor Volkov on charges of treason and unauthorized dissemination of classified state material.

Washington, through sealed channels, expressed interest in "consultative detention" regarding material affecting U.S. national security.

A third request, less explicit but unmistakable, arrived through an intermediary tied to Israeli security services, warning of dire consequences if destabilizing narratives continued to circulate.

No one was arrested.

No one was formally accused.

But their perimeter tightened.

Two black sedans idled longer than necessary at the end of the street. A man with the posture of a military professional appeared each morning outside the bakery, pretending to examine his phone while watching the reflections in the glass.

Their internet slowed unpredictably. Their movement became visible to others in ways it had not been before.

This was not assault.

This was encirclement.

And then, the pulse.

It happened at dusk, when Zurich was suspended between office-hour fatigue and evening elegance. Daniela had insisted on stepping out alone, despite Volkov's warning. She refused to live like a fugitive without first being declared one. The cold air sharpened her thoughts as she crossed Münsterbrücke, the Grossmünster's twin towers rising behind her like watchful sentinels.

She sensed it before she saw it, the subtle triangulation of movement. A van door sliding open too smoothly. Footsteps aligning behind her cadence.

A gloved hand seized her elbow with professional precision, steering rather than grabbing. Another figure moved to intercept her forward path. The choreography was practiced, economical, extraction, not execution.

Daniela twisted instinctively, but the grip tightened. The van's interior was dark, engineered for anonymity. A cloth emerged from nowhere.

And then the geometry broke.

A second vehicle, nondescript and white, mounted the curb at an angle that made no civilian sense.

Doors opened before the wheels stopped. The men who emerged did not shout either. They moved with disciplined certainty, weapons visible but unspectacular, their presence radiating authority rather than aggression.

Words were exchanged in rapid Mandarin and clipped English. The abductors recalculated in seconds.

One withdrew first. The other hesitated, too long, and found himself facing three barrels and a diplomatic immunity badge held just close enough to register.

The van retreated into traffic as though it had never existed.

Daniela stood trembling, breath fractured, eyes wide with the realization that her life had just been converted into leverage.

The men who had intervened did not introduce themselves as intelligence officers. They identified themselves as security personnel attached to the Chinese embassy. Their explanation was spare: "There are destabilizing actors in the area. We are ensuring safety of regional partners."

Regional partners.

Namibia had recently announced expanded strategic energy alignment with Beijing. Billions were at stake. Infrastructure, rare earth supply chains, maritime routes. A Namibian presidency that had defied Western pressure had become an axis of Chinese long-term planning.

Daniela was no longer merely a journalist. She was attached, emotionally and politically to that axis.

China had not saved Daniela, they saved their economic stability.

Back at the apartment, Volkov listened without interruption. His face did not display fear. It displayed confirmation. "They wanted you alive," he said quietly. "Alive is more useful than dead."

"And who does?" she asked.

He did not answer.

Because the answer was not singular.

Quietly the group gathered around the table, danger and violence has touch them again. Too many players in the shadow gunning for them. It was time to make decisions and try to keep all alive, including Daniela's family.

During the day Daniela arranged transport and accommodation for them at a UN refugee shelter in Spain.

Let them disappear into the wood-work, away from danger.

The next morning Daniela greeted her mother and brother, as she put them on the Zurich to Madrid train.

Orion had motive. So did hardliners in Moscow seeking to force his return. So did actors who believed pressure on Zurich would accelerate the Persian timeline.

In a fractured world, operations became collaborative without coordination and often innocents pay the price.

The following morning, the Russian emissary quietly arrived.

He did not come with guards. He did not come with threats. He came alone, impeccably dressed, carrying a diplomatic passport and the faint weariness of a man who had spent decades inside rooms where maps were redrawn.

The emissary's voice was measured, almost reflective.

"Moscow is divided," he said. "You are both liability and asset. Some wish you silenced. Others believe you are proof that Russia still produces men who understand global geopolitics could go both ways."

Volkov allowed himself the smallest smile.

"You are offered return," the emissary continued. "Controlled narrative. Quiet rehabilitation. You would advise from distance. No prison."

"And if I decline?"

The man's gaze shifted briefly to Daniela. "Then Moscow might be able to not protect you from those who prefer closure."

It was not a threat. It was information.

Volkov looked at the snow beginning to gather along the riverbanks. "I will not return," he said. "But I will never oppose Russia."

The emissary studied him for a long moment. "Then choose carefully where you stand." And as he turn away, he pause for a moment, and quietly whisper. "The old Colonel in the Kremlin send greetings."

Volkov just look at him, and nod once.

Far from Zurich, the Persian Pivot moved from theory into tremor.

In the Strait of Hormuz, a U.S. destroyer operating alongside allied vessels misidentified a fast-moving drone swarm during heightened alert status.

Simultaneously, Iranian radar systems registered anomalous signals suggesting missile lock patterns. Within minutes, a defensive electronic countermeasure discharged, blinding sensors on both sides.

A single interceptor missile launched prematurely, not striking a target, but detonating close enough to an Iranian patrol vessel to register as aggression.

The footage circulated before official statements could catch it.

Markets convulsed. Brent crude spiked. Insurance rates for Gulf shipping doubled overnight. Commentators invoked 1914 and sleepwalking.

As naval commanders struggled to de-escalate, a parallel event unfolded in the digital sphere.

A sophisticated cyber intrusion penetrated a regional EU energy exchange platform, manipulating trade data to simulate supply disruption.

The breach bore hallmarks of multiple signatures, Israeli, Russian, perhaps even American, but none conclusively, as attribution fractured into accusations.

For forty-eight hours, the world leaned toward catastrophe.

Israeli defence officials framed the naval incident as proof that Iran had crossed red lines. Limited strikes on proxy installations in Syria followed within twenty-four hours, framed as deterrence.

Tehran mobilized reserves. Russian reconnaissance aircraft increased patrol frequency. Chinese naval units repositioned under the language of "protecting maritime commerce."

The Persian Pivot had manifested. Not as grand declaration.

But as overlapping miscalculations.

In Washington, the US President addressed the nation with an uncharacteristically tempered tone.

The normal blustering rhetoric of dominance softened into language of recalibration. Energy security was reframed as multilateral responsibility. Corporate-state entanglements were publicly distanced. The speech did not admit error. But it shifted doctrine.

Behind the podium, fracture had become visible. Intelligence officials contradicted defence hawks in subtle briefings.

Former generals wrote op-eds warning against escalation. Markets, not morality, began influencing tempo.

And then, from Windhoek, came an unexpected transmission.

President Nadi-Omunongo called for an emergency multipolar summit under African Union auspices, hosted on Namibian soil.

The agenda was precise:
Maritime de-confliction, cyber transparency protocols, energy stability guarantees.

It was audacious, nearly improbable.

China endorsed it within hours, as Russia confirmed attendance shortly after.

Washington, after internal debate, agreed to send a senior delegation.

Israel hesitated and withdraw.

Iran signalled cautious participation.

Zurich watched as the invitation transformed Namibia from a tiny peripheral voice to a calm mature voice in Africa.

For the first time since the naval misfire, global markets steadied.

Back in the apartment overlooking the Limmat River, the group sat together in exhausted quiet.

The attempt on Daniela had left no physical scars, but it had erased any illusion of return to anonymity.

Andriy would never go home. Olena would carry the memory of exile as inheritance. Volkov had severed himself permanently from Moscow's corridors. Daniela understood that journalism had given way to consequence. Elena quietly let her thoughts drift to the peaceful red dune of the Kalahari.

Outside, snow fell softly over a city that pretended neutrality even as it hosted converging pressures.

The world had not ended. The missiles had not flown. But direction had changed.

And each of them had paid for that change with something irretrievable.

Volkov looked at Daniela, not as an asset, not as leverage, but as a future.

Namibia was no longer abstraction. It was possibility. Sanctuary not from danger, but from perpetual manipulation. A place where a future and peace might be rebuilt differently.

Far away, fleets held position, as satellites watched. Markets recalibrated.

And history, for once, paused, not because power relented, but because exposure altered incentive.

The tightening net had not suffocated them, it had forced the world to breathe differently.

Chapter 33: The Summit of Quiet Fires

The wide open Namibian skies had a way of humbling ambition.

When the delegations arrived in Windhoek, stepping from aircraft emblazoned with flags that had shaped centuries of history, they were met not by spectacle but by space. The air felt wider. The light harsher. The distances between buildings and hills and horizon impossible to ignore. It was difficult to posture under such sky. Difficult to pretend permanence.

The summit convened at State House under measured security, not militarized, but professional and alert.

Chinese advisors moved with discreet precision. American diplomats carried the visible fatigue of a system recalibrating itself in real time. Russian envoys observed more than they spoke. Iran's representatives carried the stillness of a nation that had survived isolation and decades of war before.

The Namibian President opened the proceedings without accusation.

"The world," she said calmly, "is too interconnected for hidden pivots. When energy becomes weapon, and information becomes trigger, small nations feel the tremor first. Namibia will not choose sides. Namibia will choose stability, truth and peace. The time of colonial empires and the games they play are over"

The language was deliberate. Stability. Not victory.

Behind closed doors, the tone sharpened.

The naval misfire in the Strait of Hormuz was dissected in technical detail. De-confliction corridors were negotiated. Maritime signalling protocols were updated. Cyber intrusion attribution mechanisms were drafted, imperfect, contested, but real. Energy supply chains were discussed not as instruments of dominance but as shared vulnerability and tool of progress.

It did not feel like peace, it felt like controlled damage.

The Persian Pivot did not collapse, but it slowed, and slowing, in geopolitics, is sometimes salvation.

In Zurich, the cost arrived without warning.

Daniela received the call at dawn. Short and to the point. Gas-line explosion at the refugee centre in Madrid. Her mother and brother are casualties. Sorry.

Volkov did not speak when she dropped the phone. He only crossed the room and caught her before she fell.

The grief that tore through her was not cinematic. It was raw, animal, disbelieving. She did not scream. She shook. Over and over she whispered that it was her fault, she made them visible, that her pursuit of truth had drawn a line back to her own home.

In that moment, geopolitics dissolved into something brutally simple.

Innocents had paid. And no summit resolution could undo it.

The message was clear without being signed: exposure carries consequence.

At the same hour, Elena Voss stood in a quiet corridor in Brussels, holding a sealed envelope.

She had agreed, reluctantly, to testify in a closed parliamentary session about Orion, about the Persian Pivot's corporate scaffolding, about the manipulation of energy corridors to justify military positioning.

Her testimony would not be public. It would not trend. But it would enter official EU record.

The cost for her was different.

She would never work in private defence again.

Doors closed politely.

Invitations ceased.

Her professional life fractured.

Yet when she spoke before the panel, her voice did not tremble.

"Western policy was shaped," she said evenly, "not by necessity, but by opportunity and greed. Greed disguised as self-entitled economic democratic right. By corporations who manipulate governments and politicians, just to control the global resources for profit. Immaterial the cost to innocent people"

The room did not applaud.

But it listened.

Volkov, too, was summoned, discreetly, indirectly. A neutral European capital. No cameras. No transcripts released. He described structural incentives, alliance leverage, and strategic manipulation. He did not accuse nations. He described systems.

When he finished, one official asked him quietly, "Why are you doing this?"

He answered with unexpected simplicity. "Because escalation benefits politicians, not citizens."

He left that room knowing he would never again return to Moscow.

Exile had become permanent.

The summit concluded with a joint communiqué that the media described as "measured progress." Markets stabilized. Naval patrols recalibrated. Cyber hotlines were established. Israel and Iran did not reconcile, but they stepped back from self-destruction. Washington reframed doctrine publicly. Beijing secured maritime guarantees without firing a shot. Moscow preserved relevance without escalation.

Namibia emerged not as hero, but as hinge in geopolitics.

And hinges, once visible, are never ignored again. Investment inquiries multiplied. Diplomatic traffic increased. Intelligence agencies recalculated

Windhoek as more than just a little state in Southern Africa.

Visibility was protection, till the shadow games start again.

Daniela could not attend the summit's closing ceremony.

She stood instead beneath a different sky, the wide, ancient quiet of Namibia, where her mother's ashes were scattered among red earth and dry grass that bent in the wind like whispered prayers. The funeral had been small. Private. The President had sent flowers. No speeches were made.

Grief hollowed her.

Her career, once driven by pursuit and defiance, felt suddenly distant. Every headline now carried weight measured not in clicks, but in consequence. She understood too clearly that truth could illuminate, and also attract predators.

Volkov remained beside her without speaking of geopolitics.

He built silence around her like shelter, and in that silence, something changed.

Not dramatic love. Not declaration, but an anchoring.

She realized that the world she had fought to expose would not disappear. But she did not have to fight it alone.

Andriy and Olena walked along the outskirts of Windhoek weeks later, marvelling at the emptiness of land that did not feel threatening.

After years of war-shadow and displacement, Namibia's vastness and friendly people felt like reprieve. No drones. No sirens. No air-raid apps, or Neo-Nazi gangs.

Olena found work quietly within a research institute analysing conflict resilience. Andriy advised on cyber defence frameworks for small-state infrastructure, lessons drawn from trauma, converted into protection.

They were not stateless, as Namibia has opened her doors to them.

Elena arrived last.

She came not as operative, not as strategist, but as something simpler, a woman whose past had burned away and who was ready to rebuild differently.

The President welcomed her privately. Conversations began about advisory roles, about designing transparency frameworks for resource governance, about preventing future pivots before they metastasized.

Behind the scenes, five individuals who had once moved through separate corridors of power now stood in proximity to something rarer: intention without domination.

One evening, as the sun bled into the Namib Desert and the sky turned the colour of old copper, Volkov stood beside Daniela on a ridge overlooking endless silence.

"The world did not end," she said quietly.

"No," he replied. "It changed direction."

"And my family?"

He did not answer with strategy.

He answered with a quiet gentle truth.

"They were not pawns. They were loved, innocents. That matters more than the games."

Tears came again, but softer now.

In the distance, Windhoek's lights flickered, fragile and steady.

The Persian Pivot had not vanished.

It lingered in institutions, in ambitions, in human appetites for leverage, greed and violence, but it had been slowed and hopefully redirected for a short while.

And in a sparsely populated country at the edge of the Atlantic, five lives reshaped by conflict chose not to dominate history, but to guide it quietly from behind the curtain.

Namibia would never again be invisible.

That was a risk and a responsibility carried by her leaders and people.

And under the vast Namibian sky, where space itself insisted on humility, they began again, not untouched by cost, but unwilling to surrender direction.

Epilogue

History will not record the Windhoek Summit as a revolution.

It recorded it as an adjustment and new alignment, and yet, for those who had stood at the edge of nuclear escalation and felt the tremor of something far darker beneath their feet, the adjustment was seismic.

Six months after the summit, Namibia felt different.

Not louder, but being watched as the new cool kid on the block.

Diplomatic aircraft now touched down at Hosea Kutako with quiet regularity. Delegations arrived not to dictate, but to consult. Energy partnerships were renegotiated with clauses written in clearer language, 'Namibian people first'.

Maritime agreements bore Namibian signatures in places they had never appeared before. Cyber defence cooperation frameworks carried the imprint of a small nation that had seen what weaponized invisibility could do.

Namibia was no longer peripheral.

It was visible, and visibility, President Nadi-Omunongo knew, was a blade with two edges.

She stood alone one evening on the terrace of State House, watching Windhoek's lights shimmer against the darkening hills. The air was cool, scented faintly with dust and distant rain. In the silence she could almost hear the machinery of geopolitics turning far beyond Namibia's borders.

Washington recalibrating doctrine under market pressure, Beijing consolidating influence through infrastructure rather than fleets, Moscow repositioning itself within constrained corridors of leverage, Israel and Iran locked in wary containment rather than open ignition.

The Persian Pivot had not disappeared, it had just slowed and slowing had saved millions of lives.

But slowing did not mean safe.

She understood that her country now occupied a hinge in a multipolar world. The summit had elevated Namibia from observer to participant. Investment would come. Intelligence attention would deepen. Influence would tempt.

The foundation for a better Namibia had been laid, but foundations demand vigilance.

She allowed herself one private admission beneath the wide African sky: small nations could shape history, if they resisted becoming instruments of someone else's design.

Daniela had learned to live with mornings again.

Grief did not vanish; it altered texture. Some days it arrived sharp and unannounced. Other days it rested like a low horizon, always present, not always dominant.

She had chosen to stay in Namibia.

Not because she was retreating from the world, but because she had seen enough of how the world behaved when unchecked.

Namibia, in its vastness and quiet insistence on space, offered something rare: peace and perspective.

Her journalism had changed. She no longer chased the velocity of breaking headlines.

Instead, she wrote long-form analyses on energy ethics, on multipolar accountability, on the cost paid by civilians when strategic ambition outran restraint. Her pieces circulated globally, thoughtful, measured, unwilling to inflame.

When critics asked why she had not returned to a major European bureau, she answered simply, "Clarity doesn't require proximity to power."

Volkov remained largely unseen. Officially, he was nowhere.

Unofficially, he advised carefully and sparingly, offering structural assessments to the President when requested, helping shape de-escalation frameworks for small-state resilience. He did not seek rehabilitation in Moscow. He did not seek validation in Washington. He had accepted exile not as punishment, but as chosen ground.

In Namibia's quiet, he found something he had not known he lacked: tranquillity with purpose

He and Daniela did not dramatize their bond. It grew in the unremarkable spaces, shared meals under open sky, conversations that drifted from geopolitics to memory, silence that required no performance. He did not rescue her from grief; he stood steady within it.

In the desert, anchors are not heavy. They are constant.

Elena Voss arrived in Windhoek without ceremony.

The woman who had once navigated corporate corridors where war and profit intersected, now walked through modest government buildings drafting transparency charters and oversight mechanisms. She helped design frameworks that made resource agreements traceable, auditable, and less susceptible to shadow leverage.

It was not glamorous work, yet it was necessary.

She understood systems intimately, their loopholes, their incentives, their quiet corruptions.

Now she applied that understanding in reverse. Some evenings she stood alone on her veranda, listening to wind move through dry grass, reflecting on the strange arc of her life: from facilitating strategic dominance to safeguarding small-state sovereignty.

Redemption was too dramatic a word, as purpose was enough.

Andriy and Olena discovered that Namibia's emptiness felt like room to breathe.

From war-scarred networks and cyber assault backdrops, they transitioned into building defensive digital architecture for vulnerable states.

Namibia's infrastructure became pilot ground for resilient systems, layered protections against intrusion, decentralized backups, rapid-response coordination protocols. Their experience, born in conflict, was transformed into prevention.

They spoke sometimes of returning to Europe.

They did not.

No longer stateless, they found belonging in contribution. In a country that had chosen transparency over intimidation, their skills were not weapons but shields.

One evening, months after the summit, the five of them gathered outside Windhoek on Elena's Kalahari farm, where the land opened into a sweep of red earth and pale grass stretching toward a horizon that refused containment.

There was no agenda. No briefing. Just meat on the braai, quiet laughter, and a shared understanding that each had crossed a threshold in different ways.

The sun lowered slowly, staining the sky in gradients of copper and violet. The desert did not rush twilight; it absorbed it.

Daniela stood slightly apart for a moment, watching the others. Elena speaking softly with Olena. Andriy pointing toward a distant star. Volkov seated on a rock, always watching and contemplative, but at ease.

This was not triumph.

It was settlement.

The world beyond Namibia remained complicated. Weak politicians would continue to gamble with narratives and human life. Power would still seek advantage. Ambition and greed would not evaporate.

But something had shifted as exposure had altered the balance.

Transparency had forced recalibration as more African nations begin to assert their voice.

And five individuals who had once moved along separate dangerous trajectories, now chose to remain where sky was wide enough to remind them of scale.

President Nadi-Omunongo joined them briefly that evening, without entourage, without speech. She looked across the land and said only, “We will build slowly here.”

Volkov nodded. “Slow is stable,” he replied with a wry smile.

The wind moved across the desert like a long exhale. Above them, the first stars emerged, distant, steady, and indifferent to human games, yet bearing witness to them all the same.

The world had not healed. It had not become just, but it had changed direction for a while.

Namibia stood bright, and precarious, under vast skies. And in that brightness, fragile yet deliberate, they chose to stay. Not to escape history or to shape it.

But to be part of this shining star in Africa, called the ‘Land of the Brave.’

Author's Note

This novel draws from publicly known geopolitical tensions and current global dynamics centred on global mineral and oil resources. It is not intended as a prediction, but as a geopolitical analysis on exploration of corrupt power, political leverage, and their consequences in an interconnected multipolar world.

www.ingramcontent.com/pod-product-compliance
Lightning Source LLC
LaVergne TN
LVHW052313200926
848481LV00002B/546
9789994593200